X-Ray Scattering Techniques for Epitaxial Oxide Thin Films

Daniel Sando · Paul G. Evans · Nagarajan Valanoor
Editors

X-Ray Scattering Techniques for Epitaxial Oxide Thin Films

Springer

Editors
Daniel Sando
School of Physical and Chemical Sciences
University of Canterbury
Christchurch, New Zealand

School of Materials Science
and Engineering
University of New South Wales Sydney
Sydney, NSW, Australia

Paul G. Evans
Department of Materials Science
and Engineering
University of Wisconsin-Madison
Madison, WI, USA

Nagarajan Valanoor
School of Materials Science
and Engineering
University of New South Wales Sydney
Sydney, NSW, Australia

ISBN 978-981-96-5944-9 ISBN 978-981-96-5945-6 (eBook)
https://doi.org/10.1007/978-981-96-5945-6

This Springer imprint is published by the registered company Springer Nature Singapore Pte Ltd.
The registered company address is: 152 Beach Road, #21-01/04 Gateway East, Singapore 189721, Singapore

Preface

Like many good things in science, this book is the result of casual discussion. The UNSW editors were—over coffee—lamenting a particular longstanding difficulty: the challenge of propagating the immense store of practical knowledge between generations of graduate student researchers. We knew that there were plenty of excellent texts that cover the physics of X-ray diffraction and related technologies. There was no one place, however, from which students and other researchers new to X-ray diffraction characterization could begin, especially a reference that could be used as a "how-to manual" to learn to perform complex thin film X-ray diffraction characterization. X-ray diffraction, particularly in the field of complex oxides, is highly useful, nuanced, and draws on a diverse range of ideas. Experts in these ideas and procedures are distributed around the world and across a wide range of research institutions. The UNSW editors then came up with this notion: what if we asked the experts in the field to write chapters on how to "do stuff"? (We were not interested in articles reviewing specific research accomplishments.) Our aim would be to describe the information that can be gleaned using particular approaches and to focus on how one obtains and understands the data. The UNSW team then approached Paul (Evans), who pointed out that this project would be a fantastic way to introduce the practical underpinnings of techniques involving time-resolved, coherent, and magnetic scattering—techniques that might otherwise not be accessible to students and newcomers to the field. One thing led to another, and the assembly of the book began!

Again, like many good things, the process of assembling this comprehensive book was long and faced several challenges. We approached a group of notable experts with the caveat their contribution had to be like a manual. The response was terrific. And then COVID struck! We had contributions from some authors—and some were promised—but the process was, again as with many good things, far from simple. After several detours, we had an enthusiastic group but were left without a publication outlet. We were almost back where we began, nearly two years after starting. We were fortunate at this critical point to meet Dr. Ramesh Premnath. It's not an exaggeration to say that Ramesh made this book possible. He championed the cause of this approach and convinced Springer to go ahead. He even suggested the concept of adding problems and questions after each chapter. So, all of us, editors and chapter authors alike, owe a massive vote of thanks to Ramesh. As the editors, we are also of course grateful for the huge effort of the individual topic contributors, who have provided outstanding chapters. The level of detail is simply mindboggling, and nearly every relevant X-ray technique applicable to oxide thin films is chronicled here.

Daniel Sando would like to acknowledge the assistance provided by George Bassett-Smith and Jensen Alcock. Nagarajan Valanoor would like to thank all his current group (UNSW Thin Film Fabrication) members—Kayla, Gordon, Mahak, Xueqing, Tiziana, Peggy and finally Moein Seyfouri. Moein is a superstar—he combed through the entire book to make sure it all was consistent. Moein: we remain indebted to you forever. Nagarajan Valanoor, Daniel Sando, and Paul Evans would like to dedicate this book to the people who took the time to explain the details of X-ray methods to them at Maryland, Harvard, Bell Labs, Wisconsin and CNRS-Thales; those mentors started this journey for us.

Christchurch, New Zealand/Sydney, Australia — Daniel Sando
Madison, WI, USA — Paul G. Evans
Sydney, Australia — Nagarajan Valanoor

Contents

Introduction

Paul G. Evans, Daniel Sando, and Nagarajan Valanoor

Methods employing the diffraction and scattering of X-rays, electromagnetic radiation with photon energies ranging from about 100 eV to 100 keV, are arguably the most important characterization tools available to materials researchers. X-ray scattering experiments provide versatile, quantitative, accurate, and precise characterization of the structure and properties of materials. In the context of the epitaxial growth of metal oxides, X-ray diffraction (XRD) is almost always the first characterization technique employed after material synthesis: a relatively straightforwardly acquired $\theta - 2\theta$ scan, for example, immediately provides valuable feedback about whether the growth process was successful or not. The last few decades have seen tremendous progress in the use of XRD to study thin-film oxides. Instrumentation allowing the rapid acquisition of X-ray reciprocal space maps—once only available at stations on the beamlines of synchrotron-radiation light sources—is now available off-the-shelf and installed at many universities and research institutes. These instruments can be used to acquire multidimensional maps of reciprocal space literally at the press of a button. Precise control of the sample orientation is a common component of these instruments, which include capabilities such as sample cradles with multiple angular and position axes. Diffractometers incorporating precise angular analyzers near the detector are also becoming increasingly common. Similarly, sample environments

P. G. Evans
Department of Materials Science and Engineering, University of Wisconsin-Madison, Madison, Wisconsin, USA
e-mail: pgevans@wisc.edu

D. Sando
MacDiarmid Institute for Advanced Materials and Nanotechnology, School of Physical and Chemical Sciences, University of Canterbury, Christchurch, New Zealand
e-mail: daniel.sando@canterbury.ac.nz

D. Sando · N. Valanoor (✉)
School of Materials Science and Engineering, University of New South Wales Sydney (UNSW), Sydney, Australia
e-mail: nagarajan@unsw.edu.au

D. Sando et al. (eds.), *X-Ray Scattering Techniques for Epitaxial Oxide Thin Films*,
https://doi.org/10.1007/978-981-96-5945-6_1

for non-ambient conditions, such as high and low temperatures, in situ electric fields, and liquid or gas ambient environments are also widely available.

Despite their ubiquity and usefulness, the understanding of XRD experiments and the nuances of experimental methods and analyses are often still transferred from person to person. As a result, advanced XRD techniques for the characterization of epitaxial thin films tend to be passed down orally from one generation of researcher to the next. This book aims not so much to break that cycle, but rather to reduce the reliance on it. Our goal was to construct a detailed and practical guide to the methods that researchers employ daily. The authors have distilled what they have learned—and incorporated into multiple research papers—into chapters that each serve as a manual for students and other researchers who are beginning to grasp how to carry out a particular XRD technique. The book is *not* a set of research review articles—in fact it is the opposite!—it provides an insight into the techniques that have underpinned those studies. Some basic prior knowledge is assumed: how X-rays are generated, familiarity with Bragg's Law and, to a certain extent, understanding of the chemistry of metal oxides. For further references on these topics the reader is directed to several excellent textbooks [1–14]. Otherwise, this book is a guide: for the novice thin film grower to help exploit the various possibilities enabled by XRD; for the intermediate user who wishes to exploit reciprocal space mapping to understand the subtle fingerprints of strain relaxation and other phenomena; finally for the expert who wants to push advanced concepts to the next level.

The book begins with laboratory-based XRD techniques. Chapter "Laboratory X-Ray Diffraction Analysis of Oxide Thin Films" introduces X-ray diffraction and describes its importance in the characterization of epitaxial thin films. A review of key terminology and concepts is included to assist the reader's understanding. The chapter is then divided into several subsections, each describing an important XRD measurement technique. Importantly within this context, several topics are covered in relation to the structural parameters, including strain and defects. The chapter has an ongoing focus on providing the reader practical procedures for running X-ray measurements and collecting reliable data. Representative data and figures serve as examples to explain what is observed and to demonstrate how the data can be interpreted. Overall, Chapter "Laboratory X-Ray Diffraction Analysis of Oxide Thin Films" ultimately serves as a general "how-to" guide for the acquisition and analysis of X-ray data of epitaxial thin film oxide materials.

Chapter "Synchrotron Techniques to Map Ferroelectric Domain Structures Including Superlattices" covers synchrotron techniques used to probe larger, nanometer scale structural features. These long-range periodic arrangements in epitaxial thin films are increasingly important areas of materials design and synthesis. Specifically, the chapter focuses on ferroic domains and superlattices (e.g., stacks of alternating ferroic and non-ferroic layers), for which studies of the structures and dynamics are limited, in part due to characterization challenges. Domain diffuse X-ray scattering is first presented as a method to extract meaningful information on domain configurations. The chapter reviews the key terms and core principles behind synchrotron techniques and guides the reader in the experimental acquisition of corresponding data. The analysis and interpretation of domain diffuse scattering data are presented in the context of 180 and 90° domain examples. Through an understanding of these case studies, the reader is encouraged to translate this knowledge to other domain structures.

Chapter "Coherence X-Ray Techniques for Nanoscale Characterization of Complex Oxides" discusses coherent X-ray scattering techniques for the characterization of nanoscale behaviour and dynamics. First, sources of coherent X-ray light, the speckle pattern generated by coherent X-ray scattering, as well as three key coherent X-ray scattering techniques are introduced. The utility of X-ray photon correlation spectroscopy (XPCS) in characterizing macroscopic behaviour while retaining nanoscale spatial sensitivity is highlighted. This leads to a focus on the method of XPCS, where experimental geometry, other experimental considerations, and the practicalities of running a XPCS experiments using synchrotron radiation or free-electron-laser X-ray sources are then explored in detail. Analysis of XPCS data to understand dynamical fluctuations is discussed, and fitting and plotting methods are presented. These are further illustrated with two recent XPCS studies on complex oxides focusing on ferroelectric domain fluctuation in $BaTiO_3$ thin films, and charge and orbital order fluctuations in magnetite. The chapter concludes by discussing how new advances in coherent scattering will enable fundamental insight into material dynamics at the ultrasmall and ultrafast frontier and explores the development of methods for obtaining nanoscale understanding into subtle structural or electronic effects that dictate material functionality.

Chapter "Resonant Hard X-Ray Scattering Probes of Electronic and Magnetic Order: A Practical Guide" introduces resonant elastic X-ray scattering (REXS), a tool for understanding the charge, orbital, and spin configurations of complex oxides. This chapter aims to allow scientists to explore electronic and magnetic order by providing a practical tutorial built upon relevant examples. First, the chapter introduces the fundamental theory of energy dispersive X-ray scattering, and how resonance effects can enable magnetic and charge signals to be detected. Then, the typical experimental apparatus of REXS is explained. The chapter then provides a detailed tutorial on the various steps for acquiring and understanding REXS data, including sample requirements. This is split into sections on magnetic, orbital, and charge ordering. Several case studies of each of these three types of ordering using REXS are discussed in detail. The chapter concludes with an outlook on how improvements to X-ray light sources could lead to the further evolution of resonant scattering techniques.

Chapter "Measuring Octahedral Rotations in Perovskite Oxide Heterostructures" details how synchrotron diffraction can be used to determine octahedral rotation patterns within perovskite oxide heterostructures. The chapter first describes how and why octahedral rotation patterns form in perovskite oxides and then introduces Glazer notation to describe them. Methods through which octahedral rotations can be changed, controlled, and exploited using epitaxial heterostructures are summarized, and this leads to the challenge of measuring octahedral rotations in thin films and superlattices—for which the strategies employed for bulk perovskites are unsuitable. The chapter then focuses on a detailed description of the measurements necessary for unambiguously determining a rotation pattern: which rotations give rise to which reflections, how structural domains can complicate these measurements, and the importance of relative reflection intensities. Examples of how previous studies have used these measurements to understand epitaxy, phase transitions, and structure–property relationships are explored. From here, the focus shifts to the quantitative determination of oxygen positions and bonding geometry, and structural refinement of a perovskite oxide. Following a detailed analysis of how diffraction peak intensities can be calculated, tables of atomic positions are provided, and the fitting process is explained. Finally, several important studies in which determination of octahedral rotation patterns has been essential are reviewed. The chapter concludes with an outlook on the future of determining octahedral rotations and extracting further information from half-order reflections.

Chapter "Synchrotron Time-Resolved X-Ray Diffraction for Piezoelectric Response of Ferroelectric Thin Films" introduces time-resolved X-ray diffraction for measuring the piezoelectric response of thin films. The chapter begins by outlining the fundamentals of piezoelectricity, focusing on ferroelectric materials. The chapter then explains the key advantages of time-resolved synchrotron XRD in detecting both intrinsic and extrinsic piezoelectric response. The chapter then explores how one should select different combinations of electric field waveform, probe strategies, and synchronization with X-rays, depending on the time scale of the response of interest. The experimental geometry, data acquisition, data analysis and plotting methods are then discussed. The chapter concludes with three representative studies which showcase how different time-resolved XRD methods have been used to study the technologically ubiquitous piezoelectric lead zirconate titanate—$Pb(Zr,Ti)O_3$ (PZT). The first case study explains how intrinsic piezoelectric responses in single domain films and nanorods of PZT can be observed using time-resolved XRD. The second presents the more complicated case of measurement of the extrinsic response and relaxation of polydomain films. Finally, the third case study explores the implementation of a pixel-array detector for synchronous measurements and how this allows evaluation of changes in the diffraction peak shape is described.

Everyone involved in preparing this book hopes that it will become a reference work for graduate students and experienced researchers alike. The techniques and analysis methods described here allow a broad number of physical properties of epitaxial thin films to be probed, ultimately bringing deeper understanding into these fascinating and technologically relevant materials.

References

1. B.E. Warren, *X-Ray Diffraction* (Dover Publications, 2012)
2. J. Als-Nielsen, D. McMorrow, *Elements of Modern X-Ray Physics* (Wiley, 2011)
3. L.H. Schwartz, *Diffraction from Materials* (Elsevier Science, 2012)
4. P.F. Fewster, *X-Ray Scattering from Semiconductors* (Imperial College Press, 2003)
5. U. Pietsch, V. Holy, T. Baumbach, *High-Resolution X-Ray Scattering: From Thin Films to Lateral Nanostructures* (Springer, 2013)
6. D.K. Bowen, B.K. Tanner, *High Resolution X-Ray Diffractometry and Topography* (Taylor & Francis, 2019)

7. A. Guinier, *X-Ray Diffraction: In Crystals, Imperfect Crystals, and Amorphous Bodies* (Dover Publications, 2013)
8. B.D. Cullity, S.R. Stock, *Elements of X-Ray Diffraction* (Pearson Education, 2014)
9. P.A. Cox, *Transition Metal Oxides: An Introduction to Their Electronic Structure and Properties* (Oxford University Press, 2010)
10. S. Maekawa, T. Tohyama, S.E. Barnes, S. Ishihara, W. Koshibae, G. Khaliullin, *Physics of Transition Metal Oxides* (Springer, 2004)
11. T. Vogt, D.J. Buttrey, *Complex Oxides: An Introduction* (World Scientific, 2019)
12. V.E. Henrich, P.A. Cox, *The Surface Science of Metal Oxides* (Cambridge University Press, 1994)
13. S.B. Ogale, T.V. Venkatesan, M. Blamire, *Functional Metal Oxides: New Science and Novel Applications* (Wiley, 2013)
14. N. Pryds, V. Esposito, *Metal Oxide-Based Thin Film Structures: Formation, Characterization and Application of Interface-Based Phenomena* (Elsevier Science, 2017)

Laboratory X-Ray Diffraction Analysis of Oxide Thin Films

Mohammad Moein Seyfouri, Paul G. Evans, Nagarajan Valanoor, and Daniel Sando

Abstract X-ray diffraction (XRD) is an indispensable characterization method for epitaxial thin films. It is used to determine crystal structure, phase purity, film thickness, epitaxial strain, interfacial structure, defect formation, and compositional uniformity. A wide range of samples can be measured, including simple layers of uniform chemistry, compositionally graded films and heterostructures, superlattices, free-standing membranes, and even amorphous complex oxide layers. XRD is reliable, non-destructive, and is the most rapid way to obtain quantitative information about a thin film sample. In this chapter, we provide a guide to standard laboratory-based XRD for epitaxial oxide thin films. We begin by introducing the instrumentation and definitions of angles and relevant concepts. We then describe the commonly applied methods of X-ray reflectivity and high-resolution diffraction, as well as the more advanced techniques for determining strain (reciprocal space mapping), texture (phi scans and pole figures), and mosaicity (rocking curves). The approach in this chapter is to provide the user with a toolkit: we explain exactly how to set up the measurement, which experimental parameters are appropriate, and how to analyse the data. The techniques described herein are just the starting point for more complex and in-depth techniques described in the other chapters of this book.

M. M. Seyfouri
X-Ray Facility, Mark Wainwright Analytical Centre, University of New South Wales Sydney, Sydney, Australia
e-mail: m.seyfouri@unsw.edu.au

M. M. Seyfouri · N. Valanoor · D. Sando (✉)
School of Materials Science and Engineering, University of New South Wales Sydney, Sydney, Australia
e-mail: daniel.sando@canterbury.ac.nz

N. Valanoor
e-mail: nagarajan@unsw.edu.au

P. G. Evans
Department of Materials Science and Engineering, University of Wisconsin-Madison, Madison, WI, USA
e-mail: pgevans@wisc.edu

D. Sando
School of Physical and Chemical Sciences, University of Canterbury, Christchurch, New Zealand

D. Sando et al. (eds.), *X-Ray Scattering Techniques for Epitaxial Oxide Thin Films*,
https://doi.org/10.1007/978-981-96-5945-6_2

Keywords Perovskite oxide heterostructures · Epitaxial thin film · High-resolution XRD · Reflectivity · Reciprocal space mapping

1 Introduction

X-ray diffraction and X-ray reflectivity are indispensable tools in the characterization of oxide thin film materials. A wide range of samples can be characterized, including layers of a single uniform composition, compositionally graded films and heterostructures, superlattices, free-standing membranes, amorphous complex oxide layers and interfaces between these structures. XRD is extremely reliable and is often the most rapid method to obtain quantitative information about the structure of a thin film sample. XRD methods are non-destructive and require no sample preparation.

Epitaxial thin films and heterostructures are layered materials in which structural parameters are closely linked to functional properties, such as ferroelectricity, magnetism, optical response, and electronic and thermal transport. The measurement of lattice parameters, defect density, average strain and strain gradients is a critical step in understanding these structure–property relationships.

Specific X-ray measurements address questions ranging from initial issues such as "Is the thin film epitaxial?" or "How thick is the film?" to "What phases are present in the sample?" through to more detailed questions about strain, interfacial structure, defect formation, and compositional uniformity. Table 1 summarizes structural parameters that can be measured using XRD. This chapter describes commonly used techniques and emphasizes practical applications. Each section covers key details including alignment procedures, data acquisition, and analysis. Understanding the basic concepts of thin film structural analysis underpins more advanced or specialized techniques employed in laboratory-based diffractometers, but also widely applied using synchrotron X-ray sources.

Table 1 Epitaxial thin film parameter and corresponding X-ray measurement

Structural parameter	X-ray measurement	Section(s)
Layer densities	X-ray reflectivity	3
Layer thicknesses	X-ray reflectivity High resolution diffraction	3 and 4
Defect concentration	Mosaicity: "rocking curve" Reciprocal space mapping	5 and 7
Lattice parameters	High resolution diffraction Reciprocal space mapping	4 and 7
Surface and interface roughness	X-ray reflectivity High-resolution diffraction	3 and 4
Epitaxial relationship with substrate	Phi scan Pole figure Reciprocal space mapping	6 and 7

To use this chapter, the reader should begin by reviewing the angle definitions and the terminology relevant to different X-ray measurements in Sect. 2. The subsequent sections are then self-contained, and each can be read independently. The reader may need to review parts of early sections to find concepts (e.g., sample alignment) that are presented only once to avoid repetition.

2 Definition of Terms, Angles, and Scan Types

The components of a thin film diffractometer include an X-ray source, X-ray optical systems in the incoming and exit X-ray beam paths, a goniometer, and a detector system. Here, we provide a brief description of these major components, the definitions of the angles describing the rotation axes, and terminologies relevant to specific measurement configurations. The descriptions here include references to the literature with much more information about components and definitions.

Laboratory-based thin film X-ray measurements used for high resolution XRD on epitaxial films often employ a Cu target using $K_{\alpha-1}$ characteristic radiation with wavelength $\lambda = 1.5406$ Å. This value will be used in most of the calculations throughout this chapter, so it is useful to remember it. Without loss of generality, other wavelengths can also be substituted in the formulations. In fact, laboratory sources employ other wavelengths, for example using higher brilliance sources produced by liquid jet targets based on the Ga K_{α} transition [1].

Epitaxial thin film analysis is generally performed using parallel beam optics. A series of specialized optical devices are used to condition the incident and diffracted X-ray beams, with the selection of the optics depending on the specific measurement. The optics include—but are not limited to—multilayer mirrors (i.e., Göbel mirrors) to parallelize the divergent X-ray beam, slits to limit the angular and axial divergence of the beam, and monochromators to narrow the range of wavelengths and further manipulate the angular divergence.

Epitaxial films have a high degree of crystalline perfection, and their diffraction peaks often have a small angular separation in relation to the underlying substrate or the X-ray reflections of other thin-film components. High-resolution X-ray diffraction (HRXRD) measurements using the multiple crystal diffractometers are required to resolve closely spaced sharp features [2]. *High resolution* indicates that the measurements use a highly collimated and monochromatic beam along with precise goniometry (i.e., the precision to which angles are measured).

High angular resolution in laboratory measurements is obtained using channel-cut crystals in the incident beam path to monochromate the X-ray beam by removing unwanted spectral components such as $K_{\alpha-2}$ and K_{β}, while also collimating the beam with a trade-off between intensity and resolution [3]. The typical optical setup for high-angular resolution includes a monochromator positioned in the primary beam path and a narrow slit or slits (0.1–1 mm in width) in the secondary beam path. The optical components serve to restrict the incident divergence and angular acceptance of the detector, respectively. A further improvement to the resolution can be obtained by placing an analyzer crystal before the detector, albeit at the cost of a significant decrease of detected intensity.

Before describing the various axes of a typical laboratory-based thin-film diffractometer, it is instructive to introduce the reciprocal lattice to explain the relationship between crystal structure and XRD conditions. The crystal lattice plane (hkl) can be represented by the reciprocal lattice vector whose direction is orthogonal to the lattice plane and length is equal to the inverse of the corresponding interplanar spacing. The complete set of vectors for all values of *h*, *k*, and *l* defines the reciprocal lattice [4].

The condition for diffraction of X-rays is obtained by considering the difference between the wavevectors of the incident ($\overrightarrow{k_{\text{in}}}$) and diffracted ($\overrightarrow{k_{\text{out}}}$) beams. These wavevectors are parallel to the direction of the corresponding X-ray beam and have a magnitude equal to the reciprocal of the wavelength, i.e., $\left|\overrightarrow{k_{\text{in}}}\right| = \frac{1}{\lambda}$.[1] The plane defined by the incident and scattered X-ray beam is called the scattering plane. The Bragg condition for generating a diffracted beam is [5]

$$\overrightarrow{k_{\text{out}}} - \vec{k_{\text{in}}} = \vec{Q} \tag{1}$$

The diffraction condition for a specific set of planes is satisfied when the scattering vector $\vec{Q}$ is equal to the corresponding reciprocal lattice vector.

Various types of units can be used when describing the reciprocal lattice. These include r.l.u. (reciprocal lattice units) for which the calculation for an arbitrary *d*-spacing d_{hkl} is reciprocal length $= \frac{\lambda}{2d_{hkl}}$. Another common unit is inverse angstroms (Å^{-1}), for which reciprocal length $= \frac{1}{d_{hkl}}$. Finally, as mentioned in Footnote 1, sometimes a factor of 2π is included in the wavevector length, meaning that the reciprocal lattice is given by reciprocal length $= \frac{2\pi}{d_{hkl}}$. These representations are discussed in further detail in Sect. 7.1.

Diffraction phenomena can be graphically understood using the Ewald sphere construction. A two-dimensional version of this construction is formed by drawing a circle with a radius equal to $\left|\overrightarrow{k_{\text{in}}}\right| = \frac{1}{\lambda}$ centered at the origin of the incident and diffracted beam wavevectors. The Ewald sphere shows the set of possible values of $\overrightarrow{k_{\text{out}}}$ and hence the range of scattering vectors that can be detected with the selected $\overrightarrow{k_{\text{in}}}$. Diffraction from a specific crystal plane occurs when the corresponding reciprocal lattice point is on the Ewald sphere.

Figure 1a shows the reciprocal lattice points in the 0*kl* plane of reciprocal space for a (001)-oriented $SrTiO_3$ substrate. $SrTiO_3$ has a simple cubic Bravais lattice with lattice parameter $a_{\text{SrTiO}_3} = 3.905$ Å. Figure 1a is drawn for a diffraction experiment using $\lambda = 1.5406$ Å. The maximum magnitude of $\vec{Q}$ is $\frac{2}{\lambda}$ which is possible when $\overrightarrow{k_{\text{in}}} = \overrightarrow{k_{\text{out}}}$. The maximum defines a semicircle centered at the origin in Fig. 1. Reflections outside this semicircle (e.g., the 006 in Fig. 1a) cannot be reached with X-rays of this wavelength. Figure 1a also considers a reflection geometry in which the

[1] An alternative representation is to include a factor of 2π, i.e., $\left|\vec{k_{in}}\right| = \frac{2\pi}{\lambda}$. In this chapter, we will use the notation without factor of 2π.

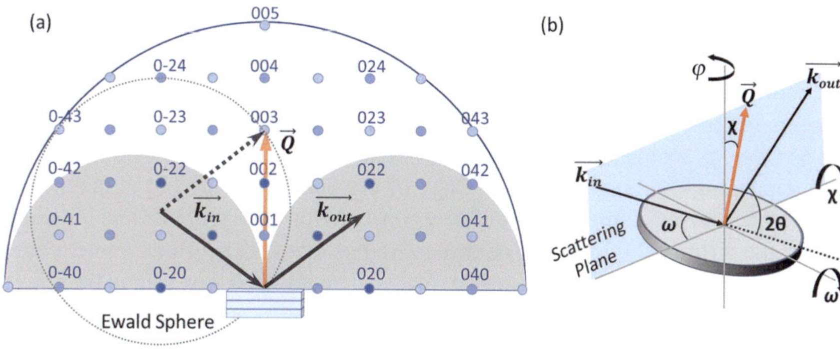

Fig. 1 **a** Section of reciprocal space for a cubic single crystal of $SrTiO_3$ with (100) orientation, including the construction of the Ewald sphere (shown as a circle) that cuts the 003 reflection. The shaded area represents the inaccessible region in the reflection geometry. **b** Schematic of four-circle goniometer

incident and outgoing beams are transmitted through the (001) surface. The shaded regions of reciprocal space cannot generate X-ray reflections for any choices of $\overrightarrow{k_{\text{out}}}$ and $\overrightarrow{k_{\text{in}}}$ in this reflection geometry, because they lie in a configuration where the sample obstructs either the incident or scattered beams.

X-ray diffractometers have a mechanical goniometer that allows the orientation of the sample to be set systematically and precisely. Several rotation axes of the goniometer are used to set the orientation of the sample with respect to the incident X-ray beam. Most commonly, the sample orientation angles are defined as an Eulerian cradle, and consist of the opening angle between the plane of the sample surface and the X-ray beam (ω), the azimuthal orientation of the sample (ϕ), and the out-of-the-diffraction plane tilt of the sample surface (χ) (Fig. 1b). The X-ray diffractometer also allows the angle of the detector with respect to the incident beam (2θ) to be selected. The direction of the scattering vector $\vec{Q}$ in reciprocal space depends on these angles. These goniometers, commonly found in laboratory-based instruments, have a total of four degrees of freedom using rotation axes (2θ, ω, χ, and φ) are known as four-circle diffractometers.

The 2θ angle measures the angle between the incident and diffracted beams and has an intuitive meaning when considering the scattering wavevector described earlier. The length of the scattering vector increases with the 2θ angle. Therefore, diffraction from lattice planes with the same orientation—but different interplanar spacing—can be observed by scanning $2\theta - \omega$. The usefulness of this interpretation will be realised for the coupled scans described in Sect. 4.1.

Different definitions of the angles setting for orienting the sample are used in other geometries. Alternatives to the Eulerian cradle design, including the kappa geometry diffractometer which uses additional sample orientation axis (κ), improve access to the scattering vectors otherwise inaccessible because of beam blockage that occurs with the traditional chi circle in Eulerian cradles [6]. Other diffractometers, such as the 6-circle diffractometers often employed in synchrotron XRD experiments, which include additional angular degrees of freedom for both the sample and the detector [7]. The calculations relating to reciprocal space and the experimental angles are more complex with additional angular degrees of freedom, [8] but are ultimately conceptually like the relationships illustrated in this section.

The motions spanned by the angles shown in Fig. 1b can be implemented differently depending on the mechanical design of the diffractometer, but in fact can be considered in terms of angles that define the orientation of the sample and the direction for the diffracted beam. Two common examples are (1) a diffractometer with a fixed X-ray source, in which the sample (ω) and detector (2θ) move about their axes, so the scattering vector for the symmetric scan rotates; (2) a diffractometer with the sample fixed horizontally while the X-ray tube and detector move about their axes, meaning that the scattering vector for a symmetric scan remains oriented in the same direction.

Out-of-plane and in-plane configurations are employed in HRXRD and investigate crystal lattice planes parallel and perpendicular to the sample surface, respectively. In the out-of-plane geometry (Fig. 2a), the direction normal to the sample lies within the scattering plane. The symmetric configuration shown in Fig. 2a has $\omega = \frac{2\theta}{2}$. In addition, asymmetric geometry allows planes inclined at a specific angle with respect to the sample surface to be probed. The angle of the planes is selected by choosing a value of the asymmetry angle α such that $\omega \pm \alpha = \frac{2\theta}{2}$. Figure 2b, c illustrates two types of asymmetric scans in grazing-exit (high angle of incidence) and grazing-incidence (low angle of incidence) diffraction geometries. We will discuss the application of each mode in more detail in Sect. 7.

XRD systems are equipped with detectors to convert the diffracted X-ray beam to electrical signals. The detectors can be categorized based on their dimensionality [3, 9]. Point detectors (known as 0D detectors) such as scintillation and proportional counters do not have positional sensitivity and the desired angular resolution for 2θ is achieved by using receiving optics like slits or an analyzer in the diffracted beam path. XRD now often includes other detector types like solid-state semiconductors in which the detecting elements are arranged in 1D or 2D arrays. Commonly, these detectors are based on heterogeneously integrated semiconductor devices termed pixel array detectors. These detectors are characterized with high speed and fast readout times and are often referred to as area detectors. Other 2D detectors often employ a gas breakdown phenomenon detected by a grid or linear array of wires

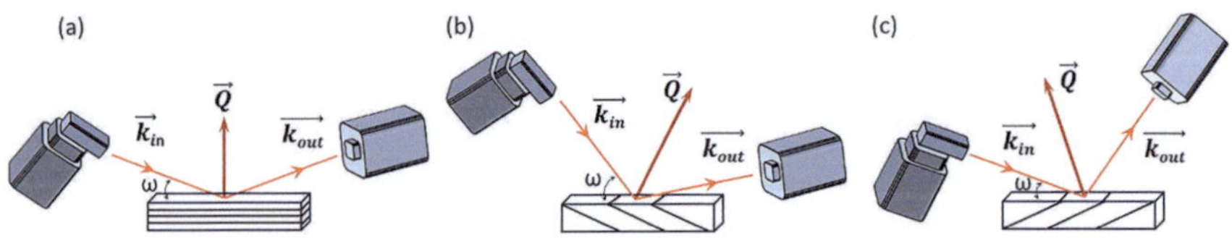

Fig. 2 **a** Symmetric out-of-plane, **b** asymmetric grazing-exit and **c** asymmetric grazing-incidence diffraction geometries

termed multiwire detectors or linear detectors. 2D detectors can be a powerful tool for studying epitaxial films through the high-speed measurement of wide areas of reciprocal space.

3 X-Ray Reflectometry (XRR)

3.1 *Introduction to X-Ray Reflectometry*

XRR employs optical reflection, with effects arising from the depth dependence of the electron density. As a result, XRR does not require diffraction from a crystalline material. XRR is applicable to a wide range of thin film samples, including crystalline (single-crystal epitaxial and polycrystalline) and amorphous thin layers. XRR profile arises from optical interference effects and can be analyzed to determine the characteristics of surface layers, such as the thickness of thin films and their density and surface/interface roughness.

X-rays incident at small angles below the critical angle (θ_c) undergo total external reflection. The value of θ_c is related to the electron density of the material through $\theta_c^2 = 2\delta$, where δ is the real part of the X-ray refractive index (which in turn depends on the electron density and the X-ray wavelength). A preliminary estimate of θ_c can be obtained by assuming that the thin film has a fixed atomic number density, a reasonable starting point for complex oxides. With this approximation, θ_c increases for materials composed of heavier elements.

For angles larger than θ_c, the incident X-ray beam penetrates through the surface and propagates within the thin film heterostructure. The propagation direction is modified by refraction, with an angular relationship set by Snell's law. The beam transmitted into the sample is partially reflected and transmitted at the surface and at each subsequent film/film and film/substrate interface due to the change in electron density at each interface. The optical amplitude and phase of each reflection depends on the thickness and composition of the layers. The propagation of the reflected beams occurs coherently and also involves reflection and transmission at each outgoing interface. The optical interference of outgoing beams produces an interference pattern that can be analyzed to obtain the structure of thin films [10–12].

In its simplest form, XRR data analysis provides a straightforward estimate of the film thickness (described in Sect. 3.3.1). In this case, the XRR measurement procedure is to vary the X-ray incident angle ω and record the intensity of X-rays in the specular reflection from the sample at angle 2θ. The required coupled scan is often termed either $2\theta - \omega$ or $\omega - 2\theta$. The reflected beam intensities often spans several orders of magnitude of dynamic range, which requires plotting on a logarithmic intensity scale. Figure 3 presents representative XRR data of a $PbZr_{0.2}Ti_{0.8}O_3$ film deposited on $SrTiO_3$ substrate and indicates how the film thickness, density, and roughness of interfaces affect the XRR profile.

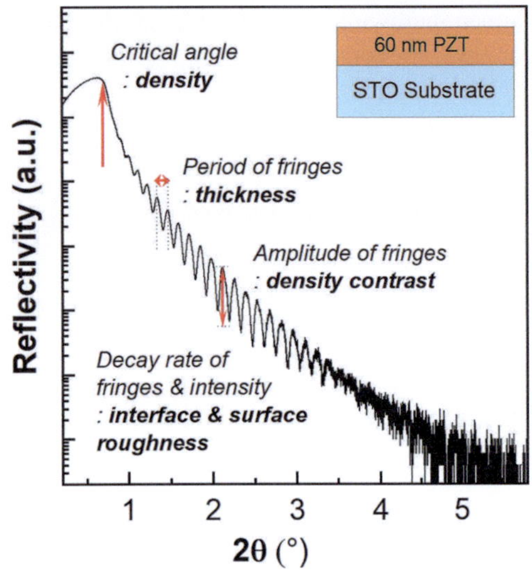

Fig. 3 XRR pattern of 60 nm (Pb,Zr)TiO_3 (PZT) film on $SrTiO_3$ (STO) substrate with surface roughness of 0.6 nm

3.2 Acquisition of X-Ray Reflectometry Data

The glancing angles involved in XRR are very small. Thus, meticulous selection of optics and precise sample alignment are highly important in obtaining data that can be analyzed reliably. We focus here on the specific example of acquiring thin-film reflectivity data using instruments based on laboratory X-ray sources operating at the wavelength provided by a sealed source employing Cu $K_{\alpha-1}$ radiation. Table 2 presents a set of initial parameters that serve as a starting point for XRR measurements of layers with thickness of up to a few 100s nm using this approach. Similar strategies apply to measurements employing other sources of X-rays, including synchrotron radiation X-ray sources.

Table 2 Parameters for XRR measurements using Cu $K_{\alpha-1}$ X-ray radiation

Typical scan parameters	• $2\theta - \omega = 0.1$–$0.2° \rightarrow 4$–$10°$ (the higher end of range is for very *thin* and/or *very smooth* films) • Step size: approximately 1/5–1/7 minimum period of oscillation (0.002–0.02°) • Time per step: 0.1–1° per min
Incident optics	• Up to 100 nm: parabolic multilayer mirror • Above 100 nm: 2 or 4-bounce monochromator depending on thickness of film
Mask (beam lateral size)	Appropriate to lateral size of sample
Incident slit	Very fine (0.05–0.2 mm) to minimize the spillover of incident beam
Detector	Pseudo 0D—very high 2θ resolution required

The first step is to select the incident and diffracted optics based on the sample size and expected thickness as mentioned in Table 2. Next is the alignment procedure which orients the sample such that the direct X-ray beam is parallel to the film surface. Precise alignment steps are also required within the reflection range to ensure the sample surface is at the center of the incident beam under grazing angle conditions. The steps are:

1. *Direct beam half cut alignment* (see **Box A**).
2. *Surface normal alignment* (see **Box B**).
3. *XRR pre-scan*: The pre-scan provides guidance for setting up the measurement conditions. It includes performing a fast $2\theta - \omega$ scan e.g., from 0.2 to ~ 8° and scan time of a few minutes. In this stage, the user should avoid $2\theta = 0$ since at this angle the detector will be subject to the direct beam and could be damaged.

If you cannot find the specular reflection (i.e., you do not see a peak during *Step 2*) one of the following may be the cause:

i. The sample is too strongly misaligned (e.g., large ω-offset): Increase the angular range of the ω-scan in *Step 2*. If you still don't see the peak, remount the sample, ensuring it lies flat on the goniometer stage.
ii. The sample is too rough to obtain a specular reflection (unlikely): → check if the sample looks mirror-like? If not, roughness might be the issue.
iii. The specular reflection is at different 2θ value: → Try values of between 0.4 and 0.8° of 2θ when performing *Step 2* above.

If all else fails, remount the sample in a different orientation, e.g., rotated by 90° in the plane.

A. Direct beam "half cut" alignment

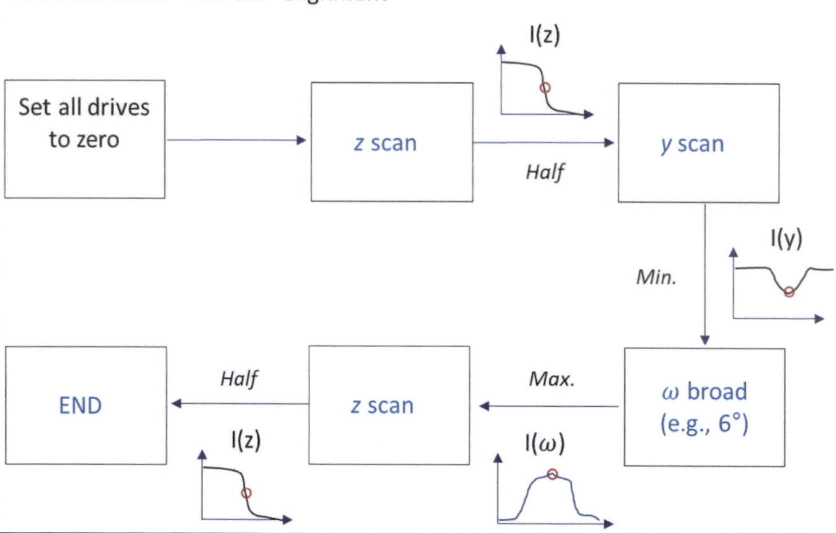

The conditions for the XRR data acquisition are tailored to the characteristics of the film and the specifications of the diffractometer. The typical angular range for XRR measurements spans from near zero (approximately 0.1°) to the angle where reflectivity intensity diminishes to the background level, with typical ranges shown in Table 2.

The step size can be determined based on the minimum oscillation period observed in an XRR pre-scan, depending on the thickness of the film. For very thin films (< 10 nm), where oscillations are broad, a step size of 0.02° is generally sufficient. However, for thicker films with fine oscillations, a smaller step size between 0.002 and 0.01° should be employed. The scanning speed (time per step) should be optimized to allow clear observation of oscillation with minimal noise, typically 0.05–1° per minute.

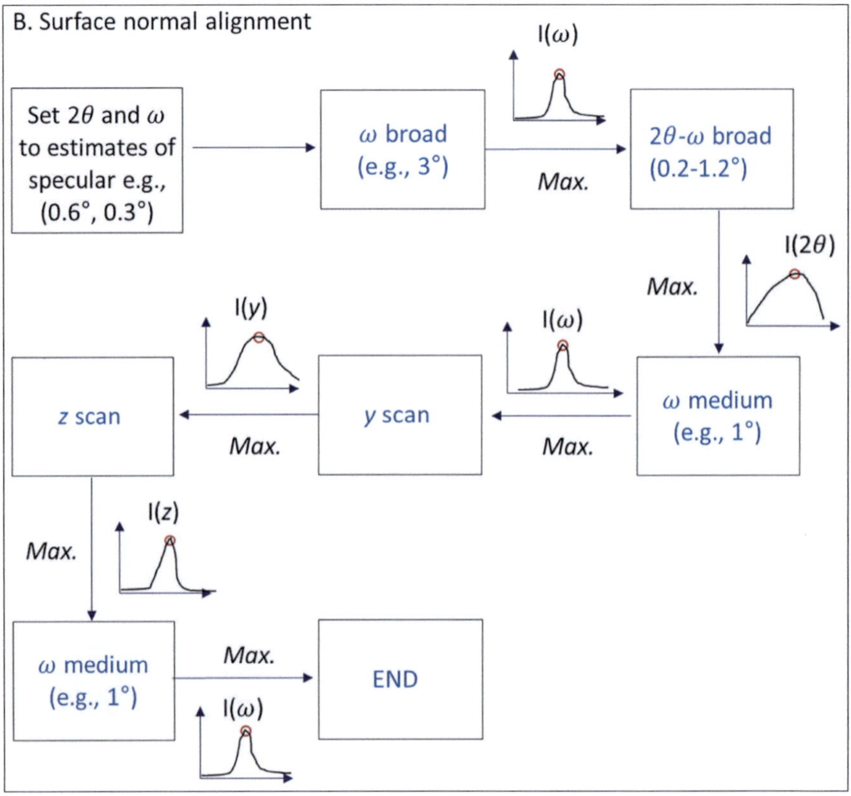

3.3 *Analysis of X-Ray Reflectometry Data*

X-ray reflectivity involves the interference of beams reflected from all the interfaces within a sample. Considering the scenario of a single layer on a substrate, a straightforward oscillation pattern is expected, as presented in Fig. 4a. After the measurement, the user can determine the critical angle and inspect the pattern for thickness fringes, which manifest as oscillations in the intensity as a function of angle. The locations of these fringes can in the first instance give an estimation of the film thickness, as described next.

3.3.1 Analysis Before Curve Fitting

The intensity oscillations in XRR patterns are known as Kiessig fringes. For a sample comprising a single layer, the oscillations arise from the incident-angle dependence of the conditions for constructive and destructive interference between reflected beams from the film/substrate interface and the surface. The angle at which a local intensity maximum occurs corresponds to the situation where the phase difference of the reflected and refracted beams is an integer factor of the wavelength. For each oscillation maximum, the angular position and film thickness are related by [13]

$$\theta_m^2 = \left(\frac{\lambda}{2t}\right)^2 m^2 + \theta_c^2 \tag{2}$$

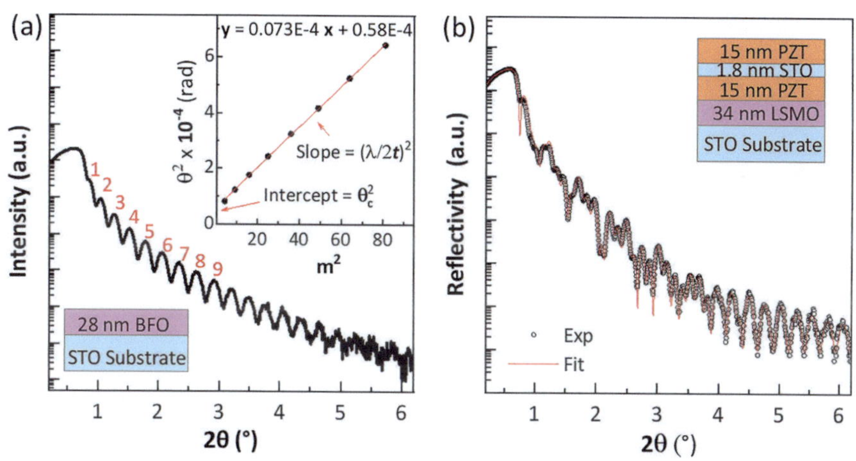

Fig. 4 **a** Thickness calculation from period of oscillation for $BiFeO_3$ (BFO)//$SrTiO_3$ (STO) sample (estimated thickness t = 28.5 nm). **b** XRR pattern of a multilayer heterostructure

where θ_m is the peak position of the mth peak, θ_c is the critical angle, and t is the film thickness. In a plot of θ_m^2 versus m^2, the film thickness and critical angle can be approximated from a linear fit to the data (inset of Fig. 4a).

From an estimate of θ_c, the real part of X-ray refractive index (δ) of the layer can be calculated as described earlier. The mass density (ρ) of a single layer is then given by [14]

$$\rho = \delta \frac{2\pi}{N_L \lambda^2 r_e} \frac{\sum_i M_i}{\sum_i (Z_i + f_i')} \tag{3}$$

This equation links the density of a layer composed of different atoms with the X-ray refractive index and composition of the layer. Here, r_e is the classical electron radius, N_L is Avogadro's number, Z_i and M_i are the atomic number and weight of the ith atom respectively, and f_i' is the atomic scattering factor of the ith atom.

The rapid decrease in intensity at angles larger than the critical angle arises from the angular dependence of the Fresnel reflectivity. An additional contribution to the decrease in the intensity is attributed to the roughness of surfaces and interfaces. For an ideally smooth surface, the XRR curve follows a decay proportional to the inverse fourth power of the scattering vector magnitude $\left(1/|\vec{Q}|\right)^4$ [15]. Surface roughness can lead to a significantly sharper decrease in intensity. The reflectivity intensity can be described using the product of the Fresnel reflectivity coefficient and a factor accounting for the root-mean-square roughness (σ). The roughness factor has a functional form similar to the Debye–Waller factor that appears in the analysis of XRD patterns acquired at a finite temperature, for which the roughness factor intensity is $\exp(-\sigma^2 |\vec{Q}|^2)$ [16].

3.3.2 Further Analysis of Reflectivity Data

Reconstructing the structure of thin films from measured data requires a more involved approach than that outlined in Sect. 3.3.1. First, a model is constructed to predict the intensity distribution as a function of angle and then the experimental X-ray reflectivity curve is compared to the theoretical profile derived from the layer structure model. The optimal values for the parameters, including thickness, density, and interface roughness, are then obtained by minimizing the residual difference between the measured and calculated reflectivity data.

For a multilayer sample, the reflectivity curve has intricate features since the intensity of the refracted wave has a complex distribution in the direction of film thickness. The complexity arises because interference involves multiple reflections at the interfaces.

The reflectivity curve can be expressed through a complex function derived from the Fresnel coefficients and is determined based on the continuity condition of the tangential components of the electric field and the conservation of tangential components of the wavevectors between adjacent layers [16]

$$R_{i,i+1} = \frac{k_{z,i} - k_{z,i+1}}{k_{z,i} + k_{z,i+1}} \tag{4}$$

Here k_z is the vertical component of wavevectors for the incoming and outgoing beams at each interface. Note that R is the fraction of the amplitude of the X-ray wave reflected at the interface. A complex value of R is possible and indicates that the phase of reflected beam is different from that of the incident beam.

Equation 4 can be used recursively to compute the intensity of the specular X-ray reflectivity. Based on the Parratt method [17], the recursion is often done self-consistently using a sum of reflection coefficients of interfaces from the bottom toward the surface. Under the assumption that the substrate is thick enough and no reflection occurs from the rear side, the square of the Fresnel coefficient at the air-film interface (involving the incident, reflected and transmitted waves at multiple interfaces within the sample) yields the specular reflectivity.

More complete analysis of the interface structure can also include more detailed descriptions of the statistical distribution of roughness and correlation of roughness between layers [15]. Combining such advanced simulation techniques, with very precise data acquisition can allow the determination of density profiles in systems for which the composition varies continuously [18].

The analysis of X-ray reflectivity data for complex structures, such as multilayers and superlattices, can be carried out using parametric modelling through simulation packages. Figure 4b shows the X-ray reflectivity pattern of a complex multilayer consisting of a $(Pb,Zr)TiO_3/SrTiO_3/(Pb,Zr)TiO_3$ structure on an $La_{0.67}Sr_{0.33}MnO_3$ buffered $SrTiO_3$ substrate. The distribution of intensity in the XRR curve in Fig. 4b is complicated because there are multiple layers contributing to the interference pattern. The layer thicknesses shown in this figure were derived from the data through a fitting process.

At the time of writing, open-source or community-developed software packages are available for the analysis of X-ray reflectivity measurements [19–21]. These packages include simulation parameters for multilayers and superlattices, as well as the capability to refine the parameters of reflectivity fits under different physical constraints.

4 High-Resolution Diffraction

4.1 Coupled $2\theta - \omega$ Scan

The coupled $2\theta - \omega$ scan is an extremely important method for the analysis of epitaxial thin films. Here, the term coupled indicates that the incident angle (after taking any offset from miscut into account) is *fixed* at half of the scattering angle (2θ). No matter what starting set of angles is selected, the $2\theta - \omega$ scan always traces a radial line in reciprocal space. Often this line is selected to be along the surface normal of the substrate or along a major crystallographic axis close to the surface normal. For the case of a highly oriented single-crystal layer, such as an epitaxial film, where the layer is in complete registry with the substrate, the coupled $2\theta - \omega$ scan gives information about the out-of-plane lattice spacing of the epilayer, that is, planes exclusively parallel to the sample surface. This scan reveals Bragg reflections from parallel planes with different interplanar spacing (e.g., 001, 002 and 003, etc.). The reader should beware, though, that epitaxial growth cannot be fully evaluated solely from observation of a single orientation in the coupled $2\theta - \omega$ scan. Additional measurements are necessary to comprehensively assess the epitaxy of a thin film layer (see Sect. 6.1).

The analysis of thick epitaxial films or films with a very low concentration of defects requires considering effects arising from dynamical diffraction. In this case, the low amplitude of scattering from each layer of atoms adds constructively and leads to diffracted beams with intensity comparable to the incident beams, effectively inducing effects similar to multiple scattering [22]. For layers with low defect densities, dynamical diffraction is important when the layer thicknesses are not much smaller than the extinction depth, which is proportional to inverse of the scattering amplitude per plane [22]. The structure of epitaxial layers with near-perfect single-crystal structures and thicknesses above approximately 10 nm requires angular resolution in the range of 0.001–0.01°, termed high-resolution diffraction earlier. With this angular resolution, the closely spaced peaks in the diffraction pattern can be resolved, allowing thicknesses, lattice parameters, and interface roughness of individual components to be determined.

4.2 Acquisition of Coupled $2\theta - \omega$ Scan Data

The first step in acquiring $2\theta - \omega$ data is sample alignment. The sample alignment includes direct beam/height alignment following the same process employed for XRR measurements (**Boxes A** and **B**). Diffraction studies of epitaxial samples also require the alignment of the crystal plane normal, sometimes termed the tilting axis adjustment. This process is summarized in **Box C**. Normal axis alignment aims to position the reciprocal lattice point corresponding to the crystallographic plane of the substrate into the scattering plane, which is the plane containing the incident and diffracted X-ray beam. The alignment process provides the precise angular settings leading to a strong reflection from the substrate. The steps are:

1. Identify the substrate reflection that exhibits the strongest intensity for your crystal and is closest to the midpoint of your intended scan range. In the case of perovskite oxides, this reflection is typically the 002_{pc} reflection. For substrates oriented differently, such as (110) or (111), the reflections to use would be the 220 and the 111, respectively. By applying Bragg's law, one can readily compute the diffraction angle associated with the chosen reflection. Many diffractometer software packages offer tools for performing these calculations.

2. Drive the 2θ axis to the calculated value for your selected reflection and set ω to half of 2θ. Detailed instructions for the alignment procedure are provided schematically in **Box C**, with an example illustrating the 002 Bragg reflection for (001)-oriented $SrTiO_3$.

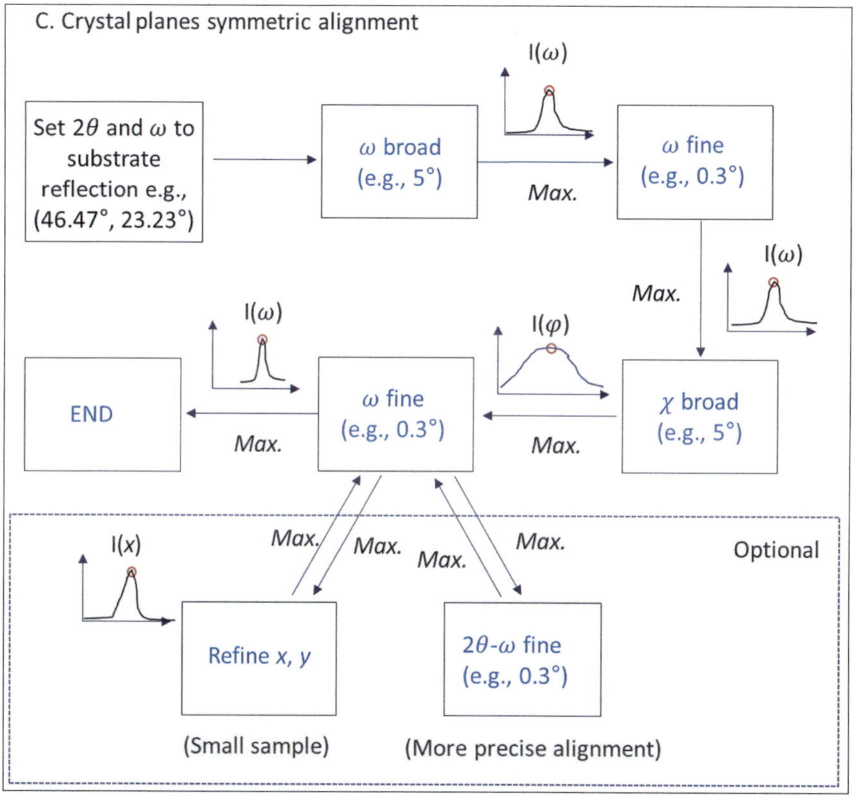

If you cannot find the substrate peak, try the following:

i. Verify the substrate crystal type and orientation.
ii. If confident in the substrate crystal orientation, conduct a ω-scan up to 12° to address potential misalignment issues. If no peak is detected, move ω back to half of 2θ, and attempt a χ scan.
iii. Increase detector counts by opening the detector, either by widening receiving slits and/or increasing active channels (for a 1D detector).
iv. Reattach the sample onto the goniometer in a different orientation. Be sure to remove silver paste or other remnants from the bottom of the sample.

The coupled $2\theta - \omega$ scan is acquired in a series of steps of the diffraction angles. When selecting the scan parameters, one must achieve a balance between intensity and resolution. The key considerations are as follows.

- HRXRD measurement usually involves scanning across a specified angular range covering a set of film-substrate Bragg reflections. Most laboratory-based 4-circle diffractometers allow access to 5–130°, sufficient to probe the primary peaks of most perovskite oxide materials.
- The choice of step size depends on the fine features present in the XRD pattern, such as thickness fringes. Initially, use a step size of 0.001–0.01° and then refine according to the required resolution.
- The scan time should be adjusted depending on the X-ray source and detector to ensure acquisition of data with a good signal-to-noise ratio. A reasonable starting point for scan speed is around 1° per minute.

4.3 Analysis of High-Resolution $2\theta - \omega$ Scan Data

4.3.1 Analysis Before Curve Fitting

In epitaxial growth, the difference in unit cell parameter (crystal structure) of the layer and underlying substrate inevitably introduces strain in the layer. The lattice matching at the film/substrate interface during crystal growth results in lattice deformation to minimize the strain energy in the layer when cooled down to room temperature. The misfit between the lattices of layer and substrate causes a separation between Bragg reflection of the epilayer and that of the substrate in the HR $2\theta - \omega$ curve. This angular separation varies with the degree of relaxation. The maximum peak separation corresponds to the pseudomorphic growth of the layer, meaning the in-plane lattice parameter of the film matches that of the substrate. Eventually, the layer partially or fully relaxes when it exceeds critical thickness. The relaxation process primarily hinges on factors such as lattice mismatch and elastic constant.

If the layer and substrate peaks do not overlap, the interplanar spacing can be inferred from the angle of the maximum intensity of the diffraction peak of the thin film layer using the Bragg equation. The lattice parameter is related to the interplanar spacing, by using $\frac{1}{d_{hkl}^2} = \frac{h^2+k^2}{a^2} + \frac{l^2}{c^2}$ assuming the (pseudo)cubic unit cell of the epilayer distorts to a tetragonal cell due to bi-axial alignment of the layer's lattice with that of the substrate.

Figure 5a depicts a high-resolution $2\theta - \omega$ scan of three epitaxial thin films of $BiFeO_3$ with different thicknesses grown on (001)-oriented $LaAlO_3$ substrates. The reflections from the $BiFeO_3$ layer appear at the low-angle side of the substrate peaks, meaning that according to Bragg's law, the epilayers have a larger lattice constant than the substrate. The out-of-plane lattice parameter of the $BiFeO_3$ film is calculated using Bragg's law to be 4.65 Å.

The low-intensity peaks flanking the primary Bragg reflections in Fig. 5a are Laue fringes. Samples with smooth interfaces and a high level of coherence (i.e., minimal strain relaxation) exhibit intensity oscillations that appear on each side of the Bragg peak. The occurrence of Laue fringes stems from a coherent scattering of a finite number of unit cells exposed to the X-ray beam. Consequently, any lattice defect that

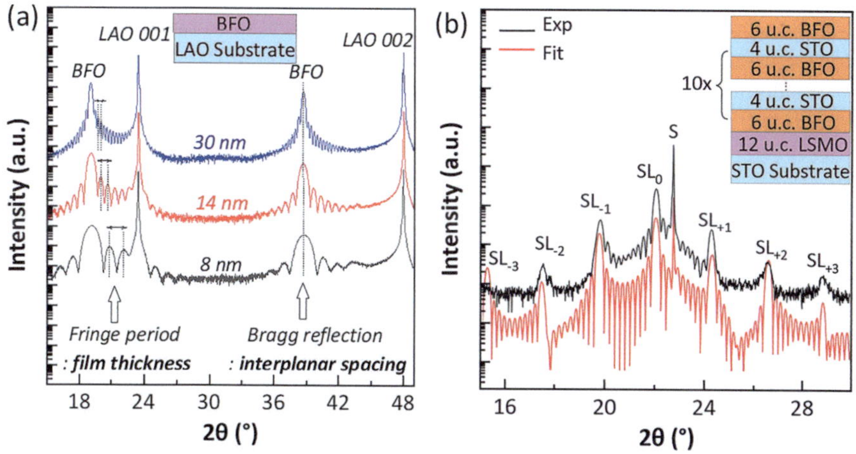

Fig. 5 **a** $2\theta - \omega$ scan for epitaxial $BiFeO_3$ (BFO) films on (001)-oriented $LaAlO_3$ (LAO). **b** X-ray diffraction pattern of a $BiFeO_3/SrTiO_3$ superlattice on $SrTiO_3$ (STO) (001)

disrupts coherence will result in complete destructive interference, rendering this method suitable for high-quality (epitaxial) films with atomically sharp interfaces. As was the case in the analysis of XRR data, the film thickness can be determined through analysis of the periodicity of the Laue fringes using [16]

$$t = \frac{\lambda}{2\Delta\omega_f \cos\theta} \tag{5}$$

where $\Delta\omega_f$ is the fringe spacing in radians, and θ is the primary diffraction peak. This estimation of layer thickness should be comparable to that obtained by XRR (see Sect. 3.3.1); however, differences in extracted thickness values are possible. Since XRR is a grazing angle technique, it probes a much larger area of the sample; whereas for high angle diffraction peaks in the $2\theta - \omega$ scan (at $2\theta = 18°$ in Fig. 5a) the higher incident angle of the X-ray beam means that it may not probe the full sample area.

4.3.2 Advanced Analysis

For comprehensive analysis, particularly for complex thin film heterostructures comprising multiple layers, advanced simulations are typically required. A quantitative fitting procedure of a HRXRD pattern can provide detailed information on thickness, composition, relaxation, and lattice parameters of an intricate epitaxial layer (such as superlattices) which is not readily accessible through basic analysis.

The term "superlattice" refers to a distinct subset of multilayers with a periodic structure, typically with atomic layers alternating between different materials. Such structures are characterized by additional diffraction peaks observed on both the low-

and high-angle sides of the film Bragg reflections. Figure 5b showcases a HRXRD scan of a ferroelectric/dielectric superlattice grown on a $SrTiO_3$ (001) substrate. These additional peaks, termed satellite or superlattice (SL) peaks, provide valuable insights into the structure. The angular position of primary superlattice reflection (0th-order) corresponds to the average lattice constant of the two layers making up the superlattice stack. Moreover, the angular separation of the satellite peaks surrounding the zeroth-order peak is proportional to the thickness of the superlattice period, that is, the thickness of repeating unit. By multiplying the superlattice period (Λ) by the number of repetitions, one can determine the total thickness of the structure:

$$\Lambda = \frac{\lambda}{2\left(\sin\theta_j - \sin\theta_{j\pm1}\right)} \tag{6}$$

Most instrument manufacturers offer proprietary software tailored for precise simulation and fitting of collected data using dynamical scattering simulation methods [23]. Moreover, there are freely available custom-made packages that calculate X-ray diffractograms of heterostructures (including superlattices) to simulate the XRD profile of epitaxial thin films on commonly used single-crystal substrates and determine the structural parameter of the layer [24]. An illustration of a simulated profile for the $BiFeO_3/SrTiO_3$ superlattice is presented in Fig. 5b.

5 *Rocking Curve* to Determine Mosaicity and Crystallographic Defect Density

The mosaic block model is a valuable intuitive framework for understanding crystallographic perfection. As shown in Fig. 6a, this model assumes a mosaic layer structure comprising numerous single crystallites that coherently scatter X-rays within their lateral and vertical dimensions, known as coherence lengths. These nanoscale blocks exhibit a slight misorientation relative to the surface normal, indicating that their crystallographic planes are not perfectly aligned but instead have a distribution of angles. To characterize the lattice distortions within this mosaic structure, tilt (out-of-plane) and twist (in-plane) angles are defined with respect to the surface normal. These angles can be quantified by analyzing rocking curves (omega scans) performed on symmetric and asymmetric (off-axis) reflections, respectively.

With rare exceptions, thin film synthesis methods yield crystals with inherent defects and imperfections. The presence of defects such as dislocations and stacking faults introduces misorientations of the crystal lattice. Rocking curve measurements offer valuable insights into the angular distribution of crystal lattice planes and are widely used to assess crystalline perfection [3, 25]. A significantly broad rocking curve implies a large distribution of the direction of the c^* lattice vector, indicative of a substantial mosaicity within the material. Conversely, an extremely narrow rocking

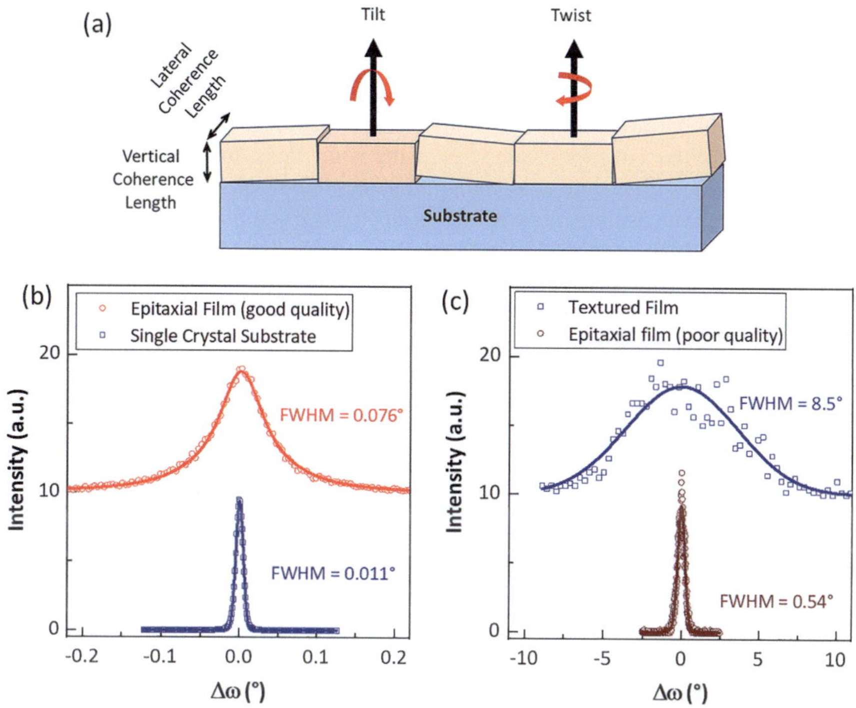

Fig. 6 **a** Mosaic block model. **b**, **c** Example rocking curves for different types of films. Rocking curves for **b** (lower) a single crystal $SrTiO_3$ substrate; (upper) a good quality epitaxial film; **c** (lower) a poor-quality epitaxial film; (upper) a textured film. Note that the single crystal has a very narrow peak width (almost perfect crystal quality), while the textured film has an extremely broad peak, reflecting a very high level of mosaicity

curve suggests minimal film mosaicity with a tightly constrained distribution of the c^* vector direction.

Before proceeding, we caution the reader about terminology: clarification is required to ensure consistency when discussing rocking curve data. The term "rocking curve scan" can refer to different concepts in the literature, especially in fields dealing with highly perfected semiconductors. For instance, a source of confusion may be the interchangeable usage of the term "rocking curve" when referring to high-resolution diffraction (coupled $2\theta - \omega$ scan) [26]. In this chapter, the term "rocking curve" is used to refer specifically to an **omega scan**.

5.1 Acquisition of X-Ray Rocking Curves

Rocking curve scans are acquired by selecting a specific Bragg reflection (e.g., 002) and fixing the 2θ angle at the peak of that reflection, i.e. fixing the *d*-spacing being probed, and then performing an omega scan. Rocking curves are typically carried out in conjunction with HRXRD (see Sect. 4.2), as such, the alignment and preparation procedures are as described in that section. To avoid coupling between misorientation and strain, rocking curve scans should be performed with somewhat relaxed resolution in the detector, for example by using wider detector slits than those used in high-resolution diffraction, or by integrating over large regions of an area detector.

Rocking curve measurements are performed on the diffraction observed in a previously measured $2\theta - \omega$ scan. The previous scan yields the appropriate 2θ angle to use for instance to probe the film peak. Due to the sample tilt when rocking, parallel beam optics are used to mitigate the defocusing effect associated with the divergent beams. The required resolution for the measurement is determined by the characteristics of the film under study. The choice of optics, for instance the type of hybrid monochromator particularly suitable for high-quality epitaxial films, establishes the maximum achievable resolution.

One can imagine that if the sample was a perfect, infinite crystal, then a rocking curve would result in a perfect Dirac delta function. Any tilt of the sample (i.e. rocking) from the perfect Bragg position would lead to a decrease in the scattered intensity. Layers with a very high degree of crystallographic perfection have an angular width primarily determined by multiple scattering effects, as described in dynamical diffraction theory [22]. For crystals with exceptional perfection and low defect concentration, a minute change in ω angle (e.g. 0.01°) will cause the diffracted intensity to drop by orders of magnitude. If the crystal has a large mosaicity, a greater deviation angle is required to move away from the ideal Bragg position before a notable decrease in the scattered intensity is observed.

Consequently, the selection of step size and counting time per step for your rocking curves will depend on the film thickness (affects diffracted intensity), film texture (influences the intensity and FWHM of the rocking curve) and sample size (impacts diffracted intensity). For instance, if the FWHM of a rocking curve is 8.5° (see Fig. 6c; top) there is very little point in performing the rocking curve with a step size of 0.01°, as this would result in *oversampling* the diffraction peak and excessively long acquisition times. On the contrary, when measuring a rocking curve of a sharp substrate reflection (e.g., Fig. 6b bottom), a step size of 0.2° would be inadequate as it would lead to *undersampling* the diffraction peak. As with the $2\theta - \omega$ scan, a good practice is to ensure that at least 10 data points above full width at half maximum of the peaks are measured, providing the first significant digit in the precision of measuring the width of the rocking curve.

5.2 Interpretation of X-Ray Rocking Curves

The width of a rocking curve depends on the intrinsic rocking curve width of a perfect crystal, instrumental broadening, mosaicity (both tilt and twist), strain at dislocations, and finally, crystal size and curvature of the sample. Typically, the intrinsic width and instrumental broadening factors are on the order of 10^{-5} rad [27], i.e., negligible for films with relatively high defect concentrations.

In heteroepitaxial layers, the effect of lattice misorientations is the primary contribution to rocking curve broadening. Due to lattice mismatch in epitaxial heterostructures, dislocations appear as strain relaxation progresses and are associated with local lattice distortions. These dislocations induce an inhomogeneous strain field in the lattice, reflected in broadening of the rocking curve peak. The density of dislocations can be estimated from the line shape of the rocking curve, and it is proportional to the square of the rocking curve width expressed by [25]

$$\rho = \frac{FWHM^2}{4.355\, b^2} \tag{7}$$

where b is the Burgers vector for different dislocation types. In the case of low dislocation density films, using the integral breadth instead of the FWHM is recommended to account for the contribution of dislocation to X-ray scattering in diffuse scattering region (tail of rocking curves). A more detailed interpretation of rocking curves necessitates fine steps in the scans, a sufficiently wide scan range, and often involves fitting with a functional form that relies on a comprehensive analysis of the mosaic model [27, 28]. In practice, the rocking curve width is usually employed as a figure-of-merit for comparing crystalline perfection among a set of epilayers, for instance when optimizing growth conditions. In general, defects in the substrate or bottom layers propagate between layers, and it is generally unusual for a thin film to exhibit higher crystalline perfection than the single-crystal substrate upon which it is grown. Exceptions can occur in cases such as the annihilation of threading dislocations. But generally, metrics such as the ratio of the FWHM of the film to the substrate FWHM are useful. These widths can be measured by performing rocking curve scans on the thin film and the substrate. For example, a ratio of close to 1 for samples grown by pulsed laser deposition (PLD) or molecular beam epitaxy (MBE) indicates excellent film quality.

Substrates for oxide thin film growth often exhibit high levels of defects. For instance, $SrTiO_3$ crystals grown by flame fusion techniques display a range of mosaic block sizes, resulting in typical rocking curve widths spanning from 0.01 to 0.05°. Crystals like $LiNbO_3$, when grown close to their stoichiometric composition from the melt, can show substantially narrower rocking curves. Advances in liquid-phase crystal growth techniques for oxide layers, exemplified by crystals grown via Czochralski method, hold promise for even narrower rocking curve widths. Notably, the rare earth scandate crystals exhibit such characteristics [29].

6 Texture Measurements and Proving Epitaxial Relationship Between Film and Substrate

In contrast to powders which contain crystal grains that are randomly oriented, thin film samples often exhibit a degree of preferred orientation, referred to as texture. This property manifests itself in the most pronounced way in epitaxial films due to the single-crystal orientation of the film's lattice planes, in turn dictated by the lattice of the substrate. The single crystal orientation in epitaxial layers is made apparent in the coupled $2\theta - \omega$ scan, where only certain Bragg reflections are observed, e.g. 001–004, or 110–330. As described above, this observation alone is not sufficient to prove that a film is epitaxial. To validate the epitaxial relationship between film and substrate, one must also determine the azimuthal (in-plane) ordering or orientation of the film.

6.1 *Phi (φ) Scan to Confirm Epitaxy*

A **φ-scan** is the most practical and convenient method for determining the film-substrate epitaxial relationship. This scan measures the intensity distribution of a specific Bragg reflection as a function of azimuthal angle by rotating the sample around its surface normal (the φ axis), while the diffraction angle 2θ (i.e., the length of the scattering vector) is fixed. By comparing the observed symmetry for film and substrate, the results of φ-scans allow us to determine the epitaxial relationship between film and substrate (see Interpretation of φ scans, below).

6.1.1 Acquisition of Phi Scan Data

Consider a simple cubic crystal. For a perfectly *00l*-oriented film, a φ-scan should be performed on a reflection from planes that are not parallel with the surface (i.e., an *asymmetric* reflection), such as h0l or hkl-type reflections. It is important to have previously measured the lattice parameter of film from a coupled symmetric scan to estimate the Bragg angle for the plane under study. The alignment procedure is

- Set the 2θ angle to the desired *d*-spacing *in symmetric mode* (i.e. $\omega = \frac{2\theta}{2}$).
- Set χ to the value corresponding to the angle between the out-of-plane vector and Bragg reflection vector (e.g., for a cubic (001) film, $\chi = 45°$ for (101) and 54.7° for (111) planes).
- Use a ω-scan to locate the substrate reflection and optimize sample alignment if necessary.

Note that signal counts might be reduced in this configuration due to the beam divergence effect in a skew-symmetric setup. Also, it can be helpful to have a slightly "open" detector to allow for possible misalignment effects in ω and χ. However, one must be mindful that if the sample has reflections that are nearby in 2θ (e.g. a film and substrate with very close lattice parameters), opening the detector reduces 2θ resolution and may introduce intensity contributions from the other reflection, thus corrupting the φ-scan accuracy. In this case, it is better to keep the detector closed, to maintain selectivity of the desired reflection.

After optimizing the peak position, a φ-scan is conducted around the full 360° range. Without removing the sample from the diffractometer, another φ-scan is performed for the film, preferably on the same plane as the substrate. Given that the film almost certainly has different lattice parameters, the 2θ will be rather different from those of the substrate.

6.1.2 Interpretation of Phi Scans

Figure 7a shows representative φ-scans of a $BiFeO_3$ thin film grown on $SrTiO_3$ (001) substrate, where both the substrate and the film show four peaks separated by 90°. The number of peaks observed in the φ-scan depends on the crystal symmetry of the film and the number of distinct in-plane orientations [3]. Notably, the peaks may not precisely appear at 0, 90, 180, 270° as the substrate may exhibit some misalignment on the goniometer. However, this minor discrepancy does not pose an issue, as the primary objective of the φ-scan is to establish the four-fold symmetry of both the substrate and film.

There are certain pitfalls to avoid during the measurement. For instance, if the sample is mounted on the stage with a substantial ω or χ offset, there is a possibility the φ-scan might miss one of the peaks. Ideally, the sample should have its normal perfectly aligned with the goniometer normal to ensure accurate results.

6.2 Pole Figure to Confirm Epitaxy

A pole figure measurement is a more detailed version of a φ-scan. Instead of acquiring data at a specific tilt angle, the sample is rotated about both φ and χ axes. It is customary to set the 2θ angle to an asymmetric reflection, such as the 101, as described above for the phi scan. One can consider a φ scan to be simply one scan of a full pole figure.

The procedure involves recording the distribution of diffraction intensity by rotating the sample around φ-axis at various tilt (χ) angles, thus constructing a polar map to depict the degree of preferred orientation within the sample. Therefore, one would expect to observe no changes in diffraction intensity for a purely randomly oriented sample. Conversely, for a fiber textured sample or a single crystal, one would expect to see a ring-shaped intensity distribution or sharp intensity spots, respectively.

Figure 7b, c presents pole figures of 111 and 002 Bragg reflections for $La_{0.67}Sr_{0.33}MnO_3$ [11] and $BiInO_3$ films [30], respectively, deposited by pulsed laser deposition on (001) and (110)-oriented $SrTiO_3$ substrates. The pole figures

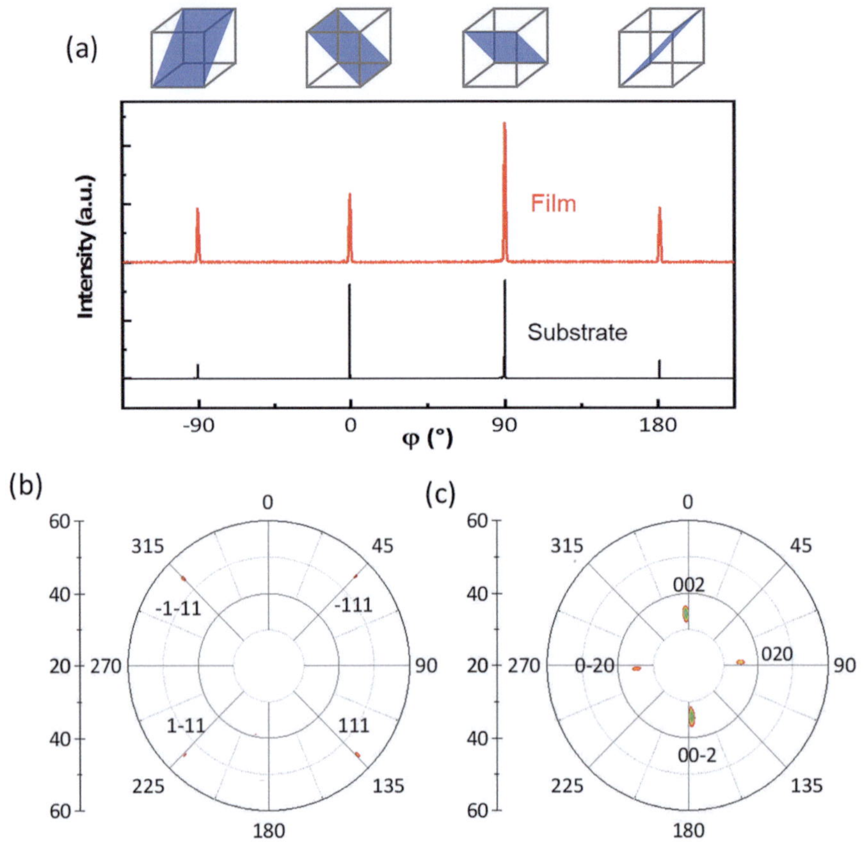

Fig. 7 **a** Phi-scan of the 101 reflection for epitaxial $BiFeO_3$//$SrTiO_3$ (001). Pole figure plots of **b** $La_{0.67}Sr_{0.33}MnO_3$//$SrTiO_3$ (001) and **c** $BiInO_3$//$SrTiO_3$ (110)

display four spots, confirming four-fold symmetry in both samples. In contrast to the sharp peaks of $La_{0.67}Sr_{0.33}MnO_3$ {111} planes, the peaks corresponding to $BiInO_3$ {002} lattice planes are noticeably broad in both radial and azimuthal directions. In pole figure measurement, the width of these peaks can serve as a metric to evaluate the degree of in-plane alignment in the crystal system [16]. Essentially, the $La_{0.67}Sr_{0.33}MnO_3$ sample shows very sharp peaks indicative of a higher degree of epitaxial quality. In contrast, the $BiInO_3$ film is grown with a much lower degree of crystalline perfection (highly textured film; not epitaxial)—as a result, the peaks are much broader.

To elucidate the epitaxial relationship further, pole figures of both the film and substrate may be superimposed within a single image. This necessitates two pole figure measurements: one for the substrate and one for the film since they will have different *d*-spacings for the chosen reflection.

7 Reciprocal Space Mapping

The insight provided by common symmetric $2\theta - \omega$ scans pertains to out-of-plane lattice planes only. Reciprocal space mapping (RSM) is used to characterize films more completely, including the in-plane lattice parameters and strain state. The RSM approach is based on thoroughly probing the reciprocal space in the vicinity of a desired reciprocal lattice point—gathering information on both the substrate and film, and specifically, their relative positions in reciprocal space [31]. RSMs measure the diffraction intensity distribution by performing a series of $2\theta - \omega$ (or 2θ, depending on the method) scans over a range of ω values (or vice versa).

Epitaxial strain usually causes the lattice parameters of the epilayer to differ from their bulk values as the layer lattice adjusts to match that of the substrate. The in-plane and out-of-plane components of an elastically distorted lattice are related based on the elasticity theory, which depends on crystal symmetry in the material. In the absence of elastic anisotropy, one can find these relationships for cubic, tetragonal, and hexagonal lattices as a function of elastic constant or the Poisson ratio [16, 25].

The precise predictions of elastic distortion of lattice constant require the determination of complex elastic tensors. RSM allows us to precisely calculate the lattice parameter of the epilayer by using the single-crystal substrate as an internal standard. It also offers unparalleled insights into the degree of strain and structural defects in epitaxial oxide films.

RSM data are most intuitively plotted using Cartesian axes representing the in-plane (Q_x) and out-of-plane (Q_z) components of the scattering vector in reciprocal space. The unit of RSM plot is typically inverse length (or *reciprocal lattice units*), expressed in reciprocal angstrom ($Å^{-1}$). The following formulae can be used to convert the angular coordinates of the diffractometer (2θ and ω) to reciprocal lattice coordinates [32],

$$Q_x = \frac{2}{\lambda} \sin\frac{2\theta}{2} \sin\left(\omega - \frac{2\theta}{2}\right), \tag{8}$$

$$Q_z = \frac{2}{\lambda} \sin\frac{2\theta}{2} \cos\left(\omega - \frac{2\theta}{2}\right). \tag{9}$$

A handy table for readily converting between various types of coordinates is provided in Table 3.

One can use RSM to probe planes that are either parallel with, or tilted relative to, the sample surface. The former is referred to as a symmetric RSM, while the latter is known as an asymmetric RSM. In general, RSMs around asymmetric reflections (e.g., 103 for a (001)-oriented film) are used to obtain information about the in-plane lattice parameter of epitaxial oxide films. However, mapping around *symmetric* reflections can indeed provide a wealth of information about domain structures, mixed phases, lattice tilt, etc. For example, if the asymmetric RSM shows multiple film peaks, or if the peak(s) appear(s) in a position that does not seem feasible based on simple strain arguments, then a symmetric RSM may shed some light. It may be that the

Table 3 Conversion between different coordinates in reciprocal space maps. The *hkl* coordinates assume a cubic lattice

From	To				
	θ, ω	r.l.u. ("raw")	r.l.u. (units of Å^{-1})	r.l.u. (units of $2\pi\text{Å}^{-1}$)	*hkl* (cubic)
θ, ω	$\times 1$	$Q_{x1} = \sin(\theta)\sin(\omega - \theta)$ $Q_{z1} = \sin(\theta)\cos(\omega - \theta)$	$Q_{x2} = \frac{2}{\lambda}\sin(\theta)\sin(\omega - \theta)$ $Q_{z2} = \frac{2}{\lambda}\sin(\theta)\cos(\omega - \theta)$	$Q_{x3} = \frac{4\pi}{\lambda}\sin(\theta)\sin(\omega - \theta)$ $Q_{z3} = \frac{4\pi}{\lambda}\sin(\theta)\cos(\omega - \theta)$	$h = \frac{2a}{\lambda}\sin(\theta)\sin(\omega - \theta)$ $l = \frac{2a}{\lambda}\sin(\theta)\cos(\omega - \theta)$
r.l.u. ("raw")	$\theta = \sin^{-1}\sqrt{Q_{x1}^2 + Q_{z1}^2}$ $\omega = \theta \pm \tan^{-1}\left(\frac{Q_{x1}}{Q_{z1}}\right)$	$\times 1$	$\times \frac{2}{\lambda}$	$\times \frac{4\pi}{\lambda}$	$h = \frac{2Q_{x1}a}{\lambda}$ $l = \frac{2Q_{z1}a}{\lambda}$
r.l.u. (units of Å^{-1})	$\theta = \sin^{-1}\left(\frac{\lambda}{2}\sqrt{Q_{x2}^2 + Q_{z2}^2}\right)$ $\omega = \theta \pm \tan^{-1}\left(\frac{Q_{x2}}{Q_{z2}}\right)$	$\times \frac{\lambda}{2}$	$\times 1$	$\times 2\pi$	$h = Q_{x2}a$ $l = Q_{z2}a$
r.l.u. (units of $2\pi\text{Å}^{-1}$)	$\theta = \sin^{-1}\left(\frac{\lambda}{4\pi}\sqrt{Q_{x3}^2 + Q_{z3}^2}\right)$ $\omega = \theta \pm \tan^{-1}\left(\frac{Q_{x3}}{Q_{z3}}\right)$	$\times \frac{\lambda}{4\pi}$	$\times \frac{1}{2\pi}$	$\times 1$	$h = \frac{Q_{x2}a}{2\pi}$ $l = \frac{Q_{z2}a}{2\pi}$
hkl (cubic)	$\theta = \sin^{-1}\left(\frac{\lambda}{2a}\sqrt{h^2 + l^2}\right)$ $\omega = \theta \pm \tan^{-1}\left(\frac{h}{l}\right)$	$Q_{x1} = \frac{h\lambda}{2a}$ $Q_{z1} = \frac{l\lambda}{2a}$	$Q_{x2} = \frac{h}{a}$ $Q_{z2} = \frac{l}{a}$	$Q_{x3} = \frac{2\pi h}{a}$ $Q_{z3} = \frac{2\pi l}{a}$	$\times 1$

λ = X-ray wavelength
a = lattice parameter along [100]

symmetric film peak splits into two components—this additional information can help decipher other confusing scans. The symmetric RSM can be thought of as an extension to the coupled $2\theta - \omega$ measurement in which the scan traces a radial line in reciprocal space to encompass reflection from planes parallel to the sample surface.

In this section, we cover the data acquisition and practical examples for both asymmetric and symmetric RSMs.

7.1 Acquisition of Reciprocal Space Mapping Data

The initial step in RSM scans includes the same alignment procedure described in Sect. 4.2. In addition, mapping around an asymmetric reflection necessitates azimuthal (φ) alignment to find the in-plane rotation of the sample that allows positioning of the desired reciprocal lattice point into the scattering plane. Detailed instructions for this alignment procedure are depicted in **Box D**. If the offset found in this step is significant, it is strongly recommended to repeat steps in **Boxes C** and **D** until ω, χ and φ are all optimized and do not change between successive scans by more than 0.1°.

The key consideration for this step is to have an approximate idea of the required sample rotation. For instance, the orientation flat (ORF) of an $SrTiO_3$ (001) substrate often aligns with the [010] direction, so to find a 113-type peak that projects along the [110] direction, the sample should be rotated in φ by 45°.

If you cannot find the peak in the φ-scan, try the following:

i. Extend the scan range to a larger value.
ii. Choose another asymmetric peak and recollect the φ-scan.
iii. If less confident about the exact reflection position, manually change φ values in small increments and collect ω-scans to find a value of φ that results in the highest peak intensity in the ω-scan. Then, follow the steps in **Box D**.

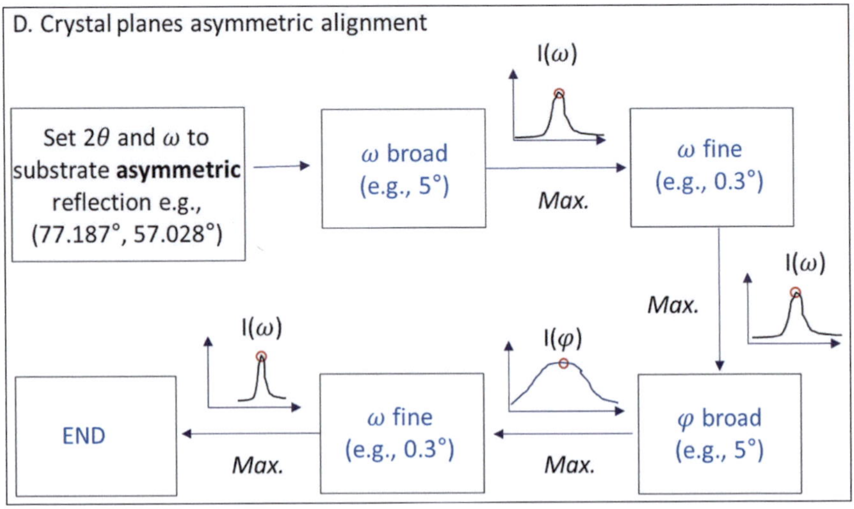

Reciprocal space mapping is a targeted measurement, meaning a specific reflection is investigated. Asymmetric RSM involves a reflection that contains an in-plane lattice component. The general idea is to take the in-plane lattice vector and add it with an out-of-plane lattice vector that gives a reflection accessible within the goniometer geometry.

Asymmetric RSM scans can be performed in either grazing-incidence or exit configurations (see Sect. 2). The X-ray penetration depth is restrained in the case of grazing-incidence setup, making it more surface sensitive (i.e. stronger film signal relative to the substrate signal) at the cost of a larger beam footprint under the lower incidence angle. Figure 8 presents RSM scans of an epitaxial high-entropy oxide film on $SrTiO_3$ (001) substrate in the two different geometries. The plots exhibit notably different shapes for grazing-incident and grazing exit scans, with the peak shape in the grazing-incidence being elongated due to divergence of the beam after diffracting from the sample.

Selecting the Reciprocal Lattice Point

Taking cubic $SrTiO_3$ (001) as an example, we can consider multiple possible asymmetric reflections from the combination of [h00] and [00l] vectors. Of all, 103, 204 and 303 are reflections of choice, since (1) using a standard 4-circle diffractometer, the lower order reflections would be inaccessible unless tilting the stage, which is not recommended (as the RSM geometry becomes more complicated); and (2) larger structure factors for these peaks lead to a higher relative intensity. The same reasoning applies to the selection of reciprocal lattice points for (pseudo) cubic perovskite substrates with (110) and (111) orientations. In this case, the typical in-plane lattice vector can be [001] and [$1\bar{1}0$] vectors, respectively, to construct the appropriate reflection [33].

For a symmetric reciprocal space map, one typically selects a high intensity reflection, such as the 002; however, it may also be beneficial to choose a reflection closest

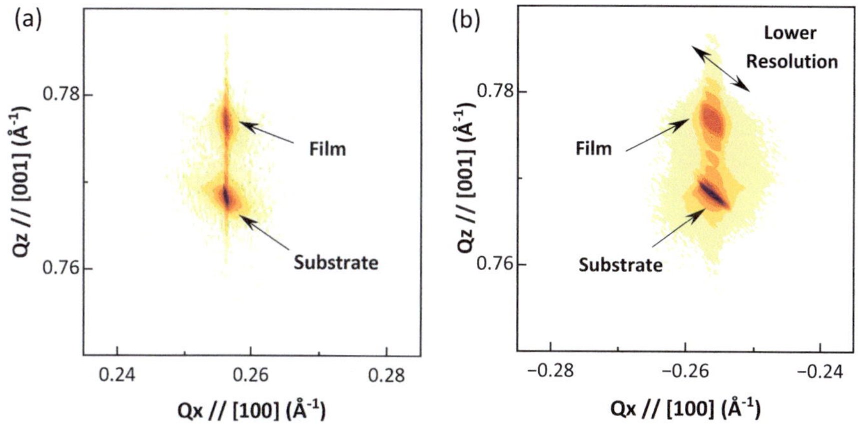

Fig. 8 Reciprocal space map scan near the 103 reflection in **a** grazing-exit and **b** grazing-incidence geometries. Note that these scans are taken with the same optics

to the asymmetric reflection of interest. For example, one could use the 003 reflection in conjunction with the 103 reflection, so that the out-of-plane (Q_z) values match, simplifying data interpretation and presentation. Note that the best angular resolution for a typical lab-based diffractometer equipped with a 2-bounce Ge (220) monochromator is at $2\theta \approx 45.2°$. Therefore, selecting Bragg reflections as close as possible to this angle will allow access to optimal resolution.

Setting Up the Scan Range for an Asymmetric RSM

The following method will help avoid the practice of 'blindly'—and inefficiently—scanning very large areas in reciprocal space. In most cases, the film peak position in reciprocal space should be between the fully strained and fully relaxed conditions, setting the upper and lower bounds on the diffractometer angles. We can calculate the expected positions for both strained and relaxed cases, along with the position of the substrate, and then set the scan range to encompass all three locations in reciprocal space. For the purposes of this exercise, let us consider 103 asymmetric reflection from a pseudocubic film with $a_{\text{film}} = 3.930$ Å epitaxially grown on $SrTiO_3$ (001) substrate ($a_{\text{sub}} = 3.905$ Å) as described next. Note that there are some cases where monoclinic splitting and/or tilting of lattice planes will shift the film peaks significantly from these calculated locations; however, in most cases, the large 'buffers' around the theoretical positions will adequately account for this peak shift.

Angular Coordinate of 103_{film} If Fully Relaxed

The Q_x (100) and Q_z (003) reciprocal lattice components are

$$Q_x = \frac{\lambda}{2(d_{100})} = \frac{1.5406}{2(3.930)} = 0.1960 \text{ r.l.u., and}$$

$$Q_z = \frac{\lambda}{2(d_{003})} = \frac{1.5406}{2(3.930/3)} = 0.5880 \text{ r.l.u.}$$

The length of the scattering vector is

$$|\tilde{Q}| = \frac{\lambda}{2 \times d_{103}} = \frac{1.5406}{2 \times \left(\frac{a_{film}}{\sqrt{10}}\right)} = 0.6198 \text{ r.l.u.},$$

or alternatively can be calculated using

$$|\tilde{Q}| = \sqrt{Q_x^2 + Q_z^2} = 0.6198 \text{ r.l.u.}$$

This value of $|\tilde{Q}|$ yields $\boldsymbol{2\theta = 76.603°}$. The off-surface angle for the asymmetric geometry is given by

$$\text{off surface angle} = \pm \,\text{a}\tan\left(\frac{Q_x}{Q_z}\right) = \pm\, 18.435°.$$

Assuming the grazing out (+) configuration, then $\boldsymbol{\omega = 56.737°}$.

***Angular Coordinate of 103**$_{substrate}$*

Using the same procedure yields the coordinates of the substrate reflection: $\boldsymbol{2\theta = 77.187°}$ and $\boldsymbol{\omega = 57.028°}$.

Angular Coordinate of 103**$_{film}$ **If Fully Strained

Find the out-of-plane lattice parameter (c_{film}) from the $2\theta - \omega$ scan, let's say this value if 4.000 Å. For simplicity, assume the film is distorted to adopt a pseudo-tetragonal unit cell, so the in-plane lattice parameters of the film are equal to that of the substrate.

The Q_x (100) and Q_z (003) reciprocal lattice components are

$$Q_x = \frac{\lambda}{2(d_{100})} = \frac{1.5406}{2(3.905)} = 0.1973 \text{ r.l.u., and}$$

$$Q_z = \frac{\lambda}{2(d_{003})} = \frac{1.5406}{2(4.000/3)} = 0.5777 \text{ r.l.u.}$$

The length of the scattering vector is

$$|\tilde{Q}| = \sqrt{Q_x^2 + Q_z^2} = 0.6105 \text{ r.l.u.}$$

This value of $|\tilde{Q}|$ yields $\boldsymbol{2\theta = 75.249°}$. The off-surface angle for the asymmetric geometry is given by

$$\text{off surface angle} = \pm \,\mathrm{a}\tan\left(\frac{Q_x}{Q_z}\right) = \pm\, 18.856°.$$

From which we calculate $\boldsymbol{\omega} = \mathbf{56.481}°$.

Calculating the Scan Range

Consider a buffer of 0.5° in both 2θ and ω for the substrate side. Depending on the film quality, consider 1.5° for both 2θ and ω on the film side. Therefore, we have the following ranges:

$\boldsymbol{2\theta}$: 73.7–78.2° (Range 4.5°; center 75.95°)
$\boldsymbol{\omega}$: 55.0–58.0° (Range 3°; center 56.6°).

7.2 *Interpretation of Reciprocal Space Mapping Plots*

Here, specific examples of 2D RSM plots shed light on the diverse capabilities of the RSM in characterizing epitaxial oxide films.

7.2.1 Shape and Position of Reciprocal Lattice Points

The reciprocal lattice points comprise information on interplanar spacing and the orientation of crystal planes determined by the length and direction of the scattering vector for a specific reflection. When performing a $2\theta - \omega$ scan with a gradually modified ω angle, the contribution of lattice tilt and strain can be separated in the 2D map since they are represented on different axes. Therefore, changes in center (position) and spread (shape) of reciprocal lattice distribution provides significant information on crystal structure, including lattice parameters and type of change in crystallinity like tilt or other distortion in lattice planes.

From the center of the reciprocal lattice point (i.e., the film peak), one can find the interplanar spacing of the corresponding plane, since the length of the reciprocal lattice vector is the inverse of *d*-spacing. If correct alignment procedures are followed, then the reciprocal coordinate vectors of the goniometer $Q_{\text{out-of-plane}}$ and $Q_{\text{in-plane}}$ will be parallel to the crystallographic axes of the sample. In this case, the in-plane and out-of-plane component of the scattering vector can be used to directly compute the interplanar spacings and thus (by considering the Miller indices of the reflection) obtain the lattice parameters. For example, if a 103_{pc} film peak is found at $Q_x = 0.2561\ \text{Å}^{-1}$, $Q_z = 0.7126\ \text{Å}^{-1}$, then the lattice parameters are given by $a_{\text{film}} = \frac{h}{Q_x} = 3.905\ \text{Å}$, and $c_{\text{film}} = \frac{l}{Q_z} = 4.210\ \text{Å}$.

Generally, the broadening of reciprocal lattice points along the $2\theta - \omega$ direction signifies the finite size of crystal domains and *d*-spacing variation due to nonuniform strain. By contrast, broadening in the ω-axis direction represents mosaicity and

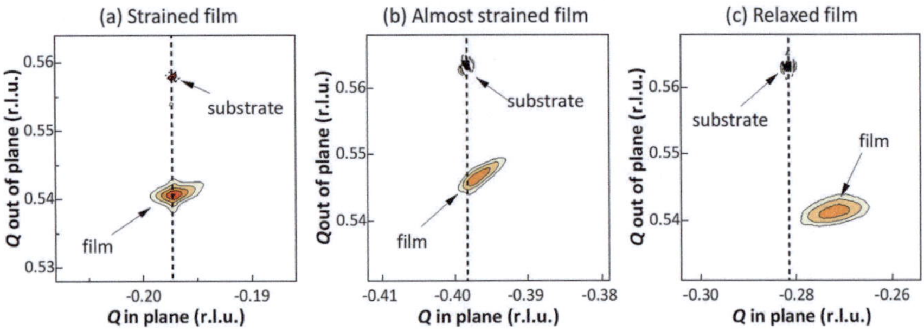

Fig. 9 Representative RSM plot of epitaxial films that are **a** strained, **b** partially relaxed and **c** fully relaxed with respect to the underlying substrate

curvature in the sample. One must be cautious however, as the impact of instrumental broadening might be significant. In addition, the relaxation (strain) state in the epilayer lattice can be inferred from the relative position of the reciprocal lattice point of the film and substrate. Let us consider a heteroepitaxial structure for a pseudocubic film on a cubic substrate in which the bulk lattice constant of the film is slightly different from that of the substrate. In this case, the film is strained to the underlying substrate if the reciprocal lattice point of the film in the 2D plot is vertically aligned (i.e., has the same Q_x component) with the substrate for the asymmetric reflections (Fig. 9a). On the other hand, if the asymmetric reflection point of the film is slightly shifted from the strained position, as shown in Fig. 9b, the film is partially relaxed or almost strained. Finally, if the reciprocal lattice point of films is along the line connecting the point to the origin of reciprocal space (that is along the direction of $2\theta - \omega$ scan) for asymmetric reflections, then the epilayer lattice is fully relaxed (Fig. 9c). In the above cases, the symmetric reciprocal lattice point of the films aligns with that of the substrate. The extreme case will be if the layer has a tilted lattice with respect to the substrate; thus, the reciprocal lattice points of the film do not line up with either symmetric or asymmetric reflection of the substrate.

7.3 Applications of Reciprocal Space Mapping to More Complex Problems

7.3.1 Unravelling Complex Layers with Overlapping Peaks in the $2\theta - \omega$ Scan

In scenarios involving a heterostructure with two layers of different materials, their out-of-plane lattice constants may become comparable if one or both layers undergo partial relaxation. This results in convoluted peaks in the coupled $2\theta - \omega$ scan, making separation challenging. However, if the two film materials exhibit a distinct difference in their bulk pseudocubic lattice constants, despite their out-of-plane lattice

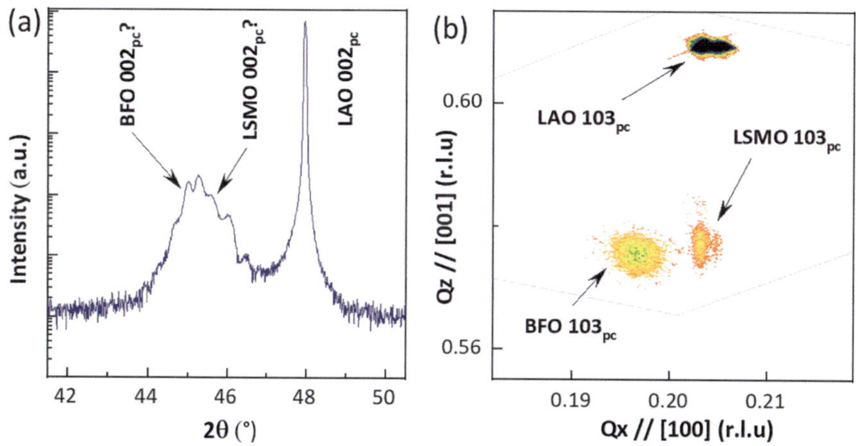

Fig. 10 Unravelling two film layers with the same out-of-plane lattice parameter. **a** Coupled $2\theta-\omega$ scan near the 002 reflection of a $BiFeO_3$ (BFO)/$La_{0.67}Sr_{0.33}MnO_3$ (LSMO)//$LaAlO_3$ (LAO) (001) heterostructure. **b** RSM near the 103_{pc} reflection for the same sample. Note that the apparent double structure in the LSMO and LAO peaks is due to the twinning in the substrate

constants being comparable in the heterostructure, their in-plane lattice constants must necessarily differ. This distinction implies that in an asymmetric RSM, the peaks of the two film layers will be separated along the Q_x axis.

Consider an example of a $BiFeO_3$/$La_{0.67}Sr_{0.33}MnO_3$//$LaAlO_3$ (001) heterostructure shown in Fig. 10a. While the substrate is easily discernible, the other two layers are less clear. They may seem to overlap, making it challenging to associate peaks with specific layers. This confusion is especially evident with apparent Laue fringes (presumably from the thinner $La_{0.67}Sr_{0.33}MnO_3$ layer) superimposed on a larger peak (presumably from the $BiFeO_3$ layer). Figure 10b demonstrates a practical approach to deconvolute these peaks through an asymmetric RSM (e.g. 103_{pc}) since the unit cell volume of $La_{0.67}Sr_{0.33}MnO_3$ and $BiFeO_3$ differs.

7.3.2 Interpretation of Monoclinic Splitting, Tilting and Superstructures in Reciprocal Space Maps

In some cases, a crystallographic tilt in the film can lead to a shift in the peak positions in reciprocal space, necessitating a more detailed analysis. There are two possible origins for horizontally split peaks in RSM around symmetric reflections. The first scenario involves the presence of two different c^* directions in the film (Fig. 11a), which physically arises from a tilting of the real space lattice parameter (either a or b) away from the substrate surface.

The second possible origin relates to the presence of a periodic structure in the film surface, such as a repeated domain structure (stripes, etc.) with a characteristic periodicity (e.g. 10–100 nm) detectable by X-rays. This situation is demonstrated

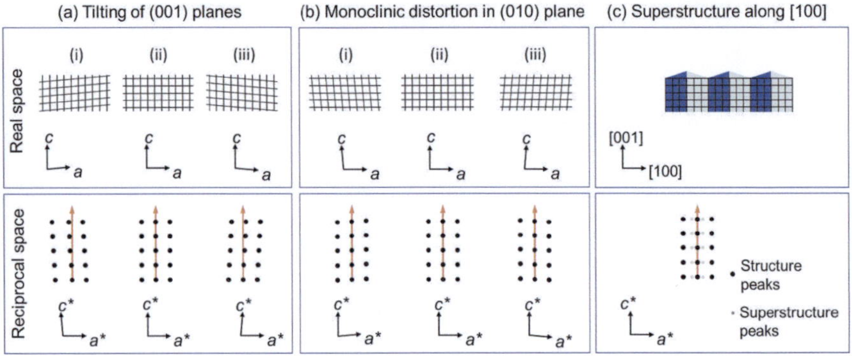

Fig. 11 Various kinds of unit cell distortions and superstructure peaks in epitaxial films. **a** Tilting of the (001) lattice planes give rise to a rotation of the c^* reciprocal lattice direction, which gives rise to a horizontal shift of the reciprocal lattice locations. **b** Monoclinic distortion, which induces a rotation of the a^* reciprocal lattice vector, and therefore shifts (only asymmetric) lattice vector positions. **c** Superstructure, coming from either ferroelectric/ferroelastic domains, or some other periodic change in density

in Fig. 11c, where superstructure peaks appear at exactly one-third of the distance between the a^* lattice length since the superstructure in this simplified case has a real space length of *three*-unit cells.

To distinguish between such cases (Fig. 11a, c), one can use symmetric scans around different diffraction orders. This approach allows us to ascertain if the magnitude of splitting depends on the diffraction order: in the first case, where the splitting in Q_x of the peak is dependent on the tilt angle of the c^* direction, going up in the diffraction order (e.g. 002) would result in an increased separation in Q_x. Conversely, if the splitting arises from a periodic structure (Fig. 11c), it will not exhibit sensitivity to the diffraction order. As a result, the splitting in the 002 reflection remains unchanged.

An example of the interpretation of crystal phases with crystallographic tilt is presented in Fig. 12. The sample is a mixed-phase $BiFeO_3$ film grown on (001) $LaAlO_3$ [34]. Due to the film thickness of approximately 80 nm, the strain relaxation process induces a complex phase mixture, resulting in multiple crystallographic phases as indicated by numerous peaks for each diffraction order (further details in [35]). While the specifics of the phase mixture are not critical here, what stands out is the linear trend in the position of the S'_{tilt} phase peaks on both the left and right sides (Fig. 12b). This observation suggests that this particular phase exhibits a crystallographic tilt of the $(001)_{pc}$ planes, cf. cases (i) and (iii) in Fig. 11a. Importantly, the absence of a superstructure is deduced from the observed increase in peak distance from Q_z with increasing diffraction order as described earlier. Note that taking the arctangent of the slope in Fig. 12b gives the tilt of the (001) planes with respect to the surface normal.

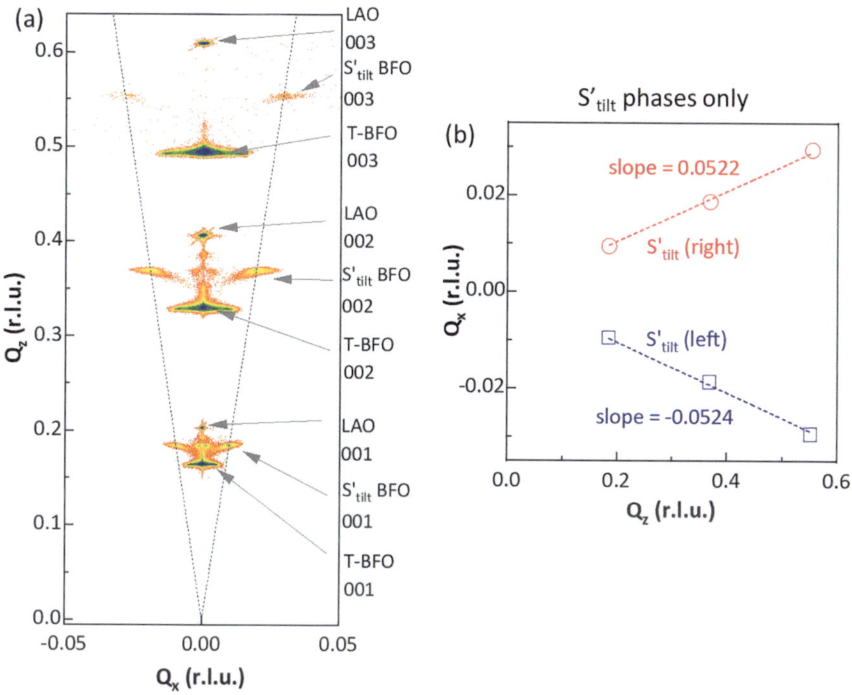

Fig. 12 Using symmetric reflections around multiple diffraction orders to determine crystallographic tilt. **a** Reciprocal space mapping around the 001, 002, and 003 reflections of $BiFeO_3$ (BFO)/ /$LaAlO_3$ (LAO) (all indices are in pseudocubic notation). Here the position of the S'_{tilt} peaks with respect to $Q_z = 0$ increases with diffraction order [case (i) and (iii) of Fig. 4a in Chap. 7]. **b** Q_x as function of Q_z for the three diffraction orders for the S'_{tilt} phase peaks. The slope of the linear trend gives a tilting angle of ~ 2.99°

There are instances where intricate tilting and monoclinic distortion angles can be observed in certain films. The situation becomes particularly complex when the distortions are oriented in different directions. For example, one may observe a monoclinic distortion in the $(110)_{pc}$ plane, coupled with a tilt of the $(001)_{pc}$ planes along the $[100]_{pc}$ direction, as shown in Fig. 13. To unravel the actual structure of the film, it is necessary to carry out a combination of asymmetric and symmetric RSMs at various azimuthal angles (φ).

8 Summary—How to Learn More

In this chapter, our focus was primarily on the practical aspects of obtaining high-quality XRD data and analyzing these data effectively. We concentrated on operational procedures, alignment techniques, data acquisition, and the nuances of data

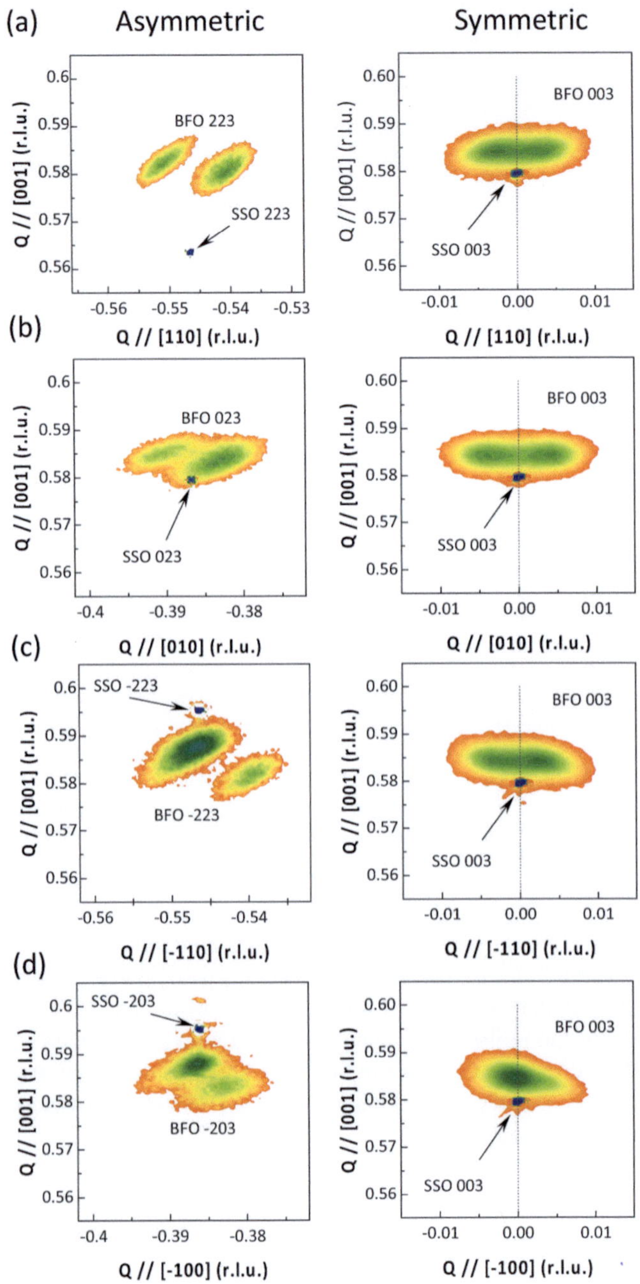

Fig. 13 Coexistence of monoclinic distortion and crystallographic tilt of the $(001)_{pc}$ planes in a $BiFeO_3//SmScO_3$ $(110)_o$ thin film (SSO = $SmScO_3$). All reflections are indexed in the pseudocubic frame. The maps were recorded at φ values of **a** − 45°, **b** 0°, **c** 45° and **d** 90°

interpretation specific to the characterization of epitaxial thin films and similar structures. For those interested in a deeper dive into the fundamental theories of X-ray scattering, and other underlying principles, we recommend consulting specialized texts such as:

- Thin film analysis by X-ray scattering [16]
- High-resolution X-ray scattering [32]
- High-resolution X-ray diffractometry and topography [36]
- X-ray and neutron dynamical diffraction: theory and applications [37]
- The series of Rigaku journal articles on "Thin film X-ray measurement techniques" [38].

The ideas in this chapter also set the stage for the development and use of a wide range of techniques, including resonant scattering, coherence, and the analysis of polarization domain structures. These are covered in other Chapters of this book.

Questions and Problems (Lab-Based X-Ray Diffraction Techniques for Oxide Thin Films)

Questions

Q 2.1. How many {101} peaks would you expect in a pole figure for an epitaxial tetragonal film?

Q 2.2. Where would the peaks appear in a pole figure for a (111)-oriented cubic film for the 312 reflection?

Q 2.3. How is it that X-ray reflectometry can be successfully used on an amorphous film?

Q 2.4. If you have several samples (with different substrates) but you do not know which is which, what X-ray diffraction method would you use to distinguish them?

Q 2.5. How can you prove "cube-on-cube" epitaxy?

Q 2.6. Explain how one could determine the miscut angle in a single-crystal substrate of a specific orientation. (Definition: miscut refers to the angular deviation of the surface orientation from the ideal crystallographic plane due to the cutting process.)

Q 2.7. How can one determine the in-plane lattice parameter of a film?

Q 2.8. Why is the rocking curve of a substrate typically so much sharper (narrower and more intense) than the rocking curve of the film?

Problems

P 2.1. Calculate the 2θ, χ, and φ angles for the {101} reflections for film and substrate for a tetragonal $BaTiO_3$ thin film ($a = 3.905$ Å, $c = 4.21$ Å) grown on $SrTiO_3$ (cubic with $a = 3.905$ Å), Assume that the substrate [100] and [010] directions correspond to $\varphi = 0$ and $\varphi = 90°$ respectively. Assume that the X-ray wavelength is $\lambda = 1.5406$ Å.

P 2.2. We have measured a series of epitaxial thin films by standard X-ray diffraction $2\theta - \omega$ scans. (The X-ray source is the Copper $K_{\alpha-1}$ line—with a wavelength of $\lambda = 1.5406$ Å.) Unfortunately, we mixed up the samples and named the data files incorrectly, so we do not know which XRD pattern belongs to which sample. We know that the possible film materials are the following [One of the samples is a heterostructure comprising two film layers]:

Film [all (001) orientation]	Bulk pseudo-cubic lattice parameter (Å)
$La_{0.67}Sr_{0.33}MnO_3$ (LSMO)	3.854
$La_{0.67}Ca_{0.33}MnO_3$ (LCMO)	3.888
$SrRuO_3$ (SRO)	3.922
$BiFeO_3$ (BFO)	3.966

The possible substrates are the following:

Substrate [all (001) orientation]	Bulk pseudo-cubic lattice parameter (Å)
$SrTiO_3$ (STO)	3.905
$NdGaO_3$ (NGO)	3.861
$LaAlO_3$ (LAO)	3.791

We know that all the films are fully strained, which means that the in-plane lattice parameters of the film are the same as the substrates upon which they are grown. For simplicity, assume that the unit cell is distorted such that *the unit cell volume is conserved.*

Below, the diffraction patterns for the four samples are shown. On the left is a wide scan showing two diffraction orders; on the right is a zoom of the range 41–49°, from which you can extract the approximate angles of the substrate and film peaks to calculate the lattice parameters.

Sample 1.

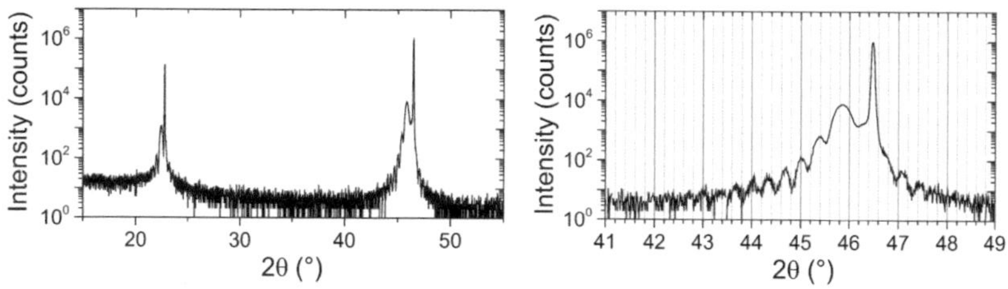

Sample 2.

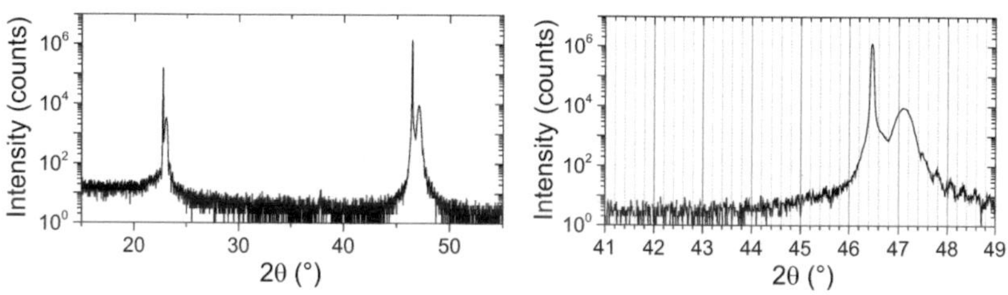

Sample 3.

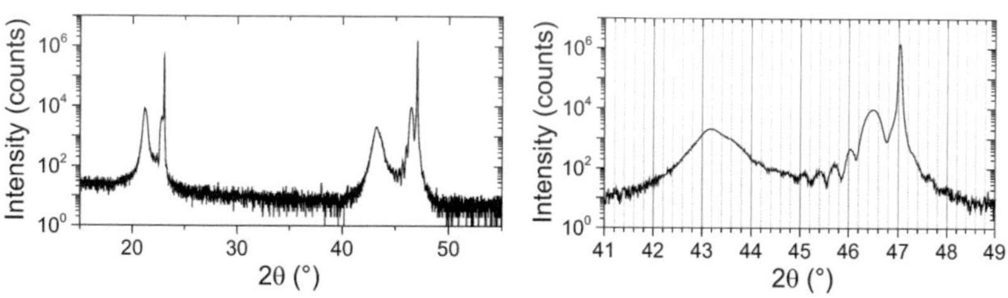

Sample 4.

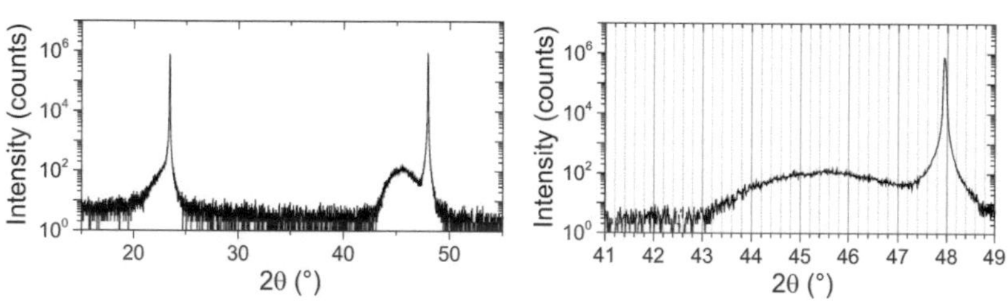

Based on these diffraction patterns, use Bragg's law and the lattice constants provided above to classify the samples, i.e., the film material and the substrate, for all four samples.

P 2.3. Explain, by calculating the corresponding 2θ and ω angles, why the 103_{pc} reflection is most commonly used for asymmetric reciprocal space mapping. (Hint: what are the goniometer angles for e.g., the 101_{pc} reflection?)

P 2.4. Many perovskite oxide thin films are grown on (001)-oriented cubic substrates, but one can also consider high-index orientations such as (210). In this case, the [210] direction is the surface normal and the in-plane directions are [001] and $[1\bar{2}0]$ respectively. Using the method and result of Problem 2.3 above as a guide, suggest several reflections of (210) $SrTiO_3$ ($a = 3.905$ Å) suitable for asymmetric reciprocal space maps conducted on a typical laboratory diffractometer. Keep in mind the limitations of the goniometer's geometry.

P 2.5. Calculate the electron density and the critical angle for $SrTiO_3$ using copper $K_{\alpha-1}$ transition as the X-ray source. Next, explore the potential effects of using longer wavelengths in X-ray reflectivity (XRR) measurements.

P 2.6. The value of Poisson's ratio is often taken as 0.5 to simplify the calculation of the lattice parameter of a strained unit cell, ensuring conservation of the unit cell volume. Demonstrate how this value is derived.

P 2.7. Determine the angles for observing asymmetric 103 reciprocal lattice point of $LaAlO_3$ (001) substrate in grazing-incident and grazing-exit geometry using the copper $K_{\alpha-1}$ transition as the X-ray source.

P 2.8. Presented below are XRD patterns ranging from the (001) to the (004) peaks for epitaxially grown single layer films: (a) BFO//LAO (001) and (b) SRO// STO (001). However, these are not plotted on a logarithmic intensity scale.

i. In terms of the intensity, qualitatively describe the trends observed in the peaks.
ii. Justify your answer to (i) by considering the structure and atomic factors. Take the example of a cubic ABO_3 oxide perovskite, where: A sits at (0, 0, 0), B sits at (½, ½, ½) and O at (½, 0, 0), (0, ½, 0) and (0, 0, ½).

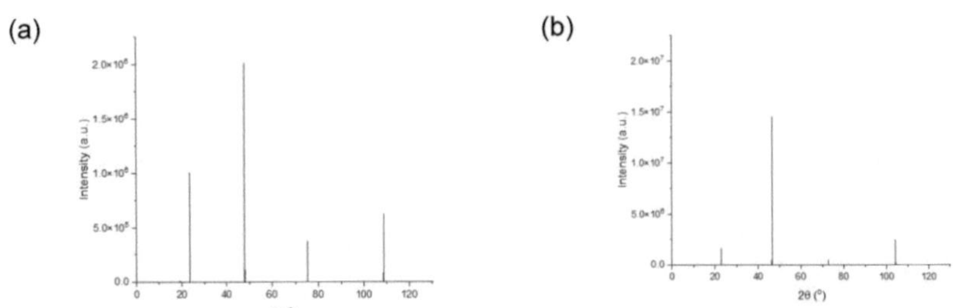

P 2.9. Bismuth Ferrite ($BiFeO_3$) is the prototypical example of a multiferroic, remaining a material of great research interest. Dependent on the epitaxial strain, $BiFeO_3$ can take on different phases: a tetragonal like T′ phase when under large compressive strain or a rhombohedral-like R′ phase when under moderate compressive or tensile strains. However, when $BiFeO_3$ films are grown past a critical thickness, strain relaxation drives the formation of a mixed phase, consisting of T′, R′ and a S′ phase (a compressively strained variant of the R′ phase). From the information available in the below figure [34]:

i. Estimate the critical thickness.
ii. Identify the S′ peak(s).
iii. Identify the K_β peak(s).
iv. Estimate the c/a ratios for the T′ and S′ phases, assuming that both are strained to the substrate ($LaAlO_3$, $a = 3.79$ Å).
v. Explain why fewer Laue oscillations are observed at higher thicknesses?

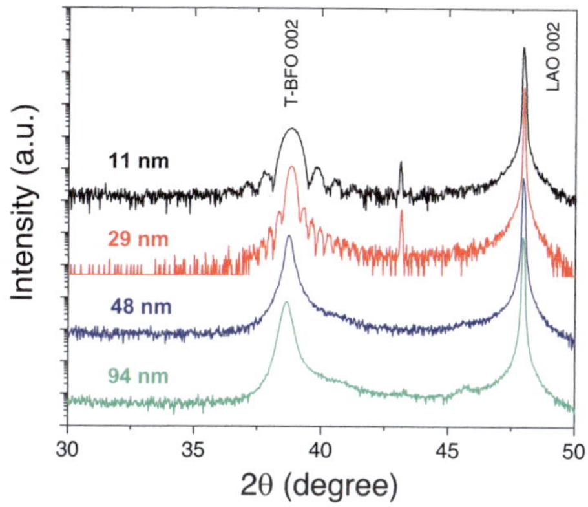

References

1. M. Otendal, T. Tuohimaa, U. Vogt, H.M. Hertz, A 9keV electron-impact liquid-gallium-jet X-ray source. Rev. Sci. Instrum. **79**, 016102 (2008)
2. P.F. Fewster, N.L. Andrew, Applications of multiple-crystal diffractometry. J. Phys. D Appl. Phys. **28**, A97 (1995)
3. G.F. Harrington, J. Santiso, Back-to-basics tutorial: X-ray diffraction of thin films. J. Electroceram. **47**, 141 (2021)
4. C. Kittel, Wave diffraction and the reciprocal lattice, in *Introduction to Solid State Physics*, 8th edn. (Wiley, 2004), pp. 23–47
5. Y. Waseda, E. Matsubara, K. Shinoda, Scattering and diffraction, in *X-Ray Diffraction Crystallography: Introduction, Examples and Solved Problems* (Springer, Berlin, Heidelberg, 2011), pp. 67–106
6. I.K. Robinson, H. Graafsma, A. Kvick, J. Linderholm, First testing of the fast kappa diffractometers at National Synchrotron Light Source and European Synchrotron Radiation Facility. Rev. Sci. Instrum. **66**, 1765 (1995)
7. H. You, Angle calculations for a '4S+2D' six-circle diffractometer. J. Appl. Cryst. **32**, 614 (1999)
8. G. Thorkildsen, R.H. Mathiesen, H.B. Larsen, Angle calculations for a six-circle κ diffractometer. J. Appl. Cryst. **32**, 5 (1999)
9. S. Kobayashi, K. Inaba, X-ray thin-film measurement techniques. VIII. Detectors and series summary. Rigaku J. **28**, 8 (2012)
10. D. Sando, C. Carrétéro, M.N. Grisolia, A. Barthélémy, V. Nagarajan, M. Bibes, Revisiting the optical band gap in epitaxial $BiFeO_3$ thin films. Adv. Opt. Mater. **6**, 1700836 (2018)
11. T. Young et al., Structural, magnetic, and ferroelectric properties of T-like cobalt-doped $BiFeO_3$ thin films. APL Mater. **6**, 026102 (2018)

12. M.M. Seyfouri, D. Zhang, C. Menéndez, D. Sando, Q. Zhang, C. Cazorla, J. Seidel, D. Wang, Engineering photoresponse in epitaxial $BiFe_{0.5}Cr_{0.5}O_3$ thin films through structural distortion. J. Phys. Chem. C **126**, 14329 (2022)
13. V. Mandić, S. Kurajica, I. Panžić, A. Bafti, J. Šipušić, K. Mužina, F. Brleković, L. Gigli, M. Gaboardi, Utilization of conventional PXRD apparatus for characterization of thin-films using reconsidered equations for XRR. Surf. Interfaces **36**, 102554 (2023)
14. M. Yasaka, X-ray thin-film measurement techniques. V. X-ray reflectivity measurement. Rigaku J. **26** (2010)
15. S.K. Sinha, E.B. Sirota, S. Garoff, H.B. Stanley, X-ray and neutron scattering from rough surfaces. Phys. Rev. B **38**, 2297 (1988)
16. M. Birkholz, Thin Film Analysis by X-Ray Scattering, 1st edn. (Wiley, 2005)
17. L.G. Parratt, Surface studies of solids by total reflection of X-rays. Phys. Rev. **95**, 359 (1954)
18. B.K. Tanner, Grazing incidence X-ray reflectivity and scattering, in *Handbook of Advanced Non-Destructive Evaluation*, ed. by N. Ida, N. Meyendorf (Springer International Publishing, Cham, 2018), pp.1–34
19. Y. Gerelli, Aurore: new software for neutron reflectivity data analysis. J. Appl. Cryst. **49**, 330 (2016)
20. M. Björck, G. Andersson, GenX: an extensible X-ray reflectivity refinement program utilizing differential evolution. J. Appl. Cryst. **40**, 1174 (2007)
21. A.R.J. Nelson, S.W. Prescott, Refnx: neutron and X-ray reflectometry analysis in Python. J. Appl. Cryst. **52**, 193 (2019)
22. B.W. Batterman, H. Cole, Dynamical diffraction of X rays by perfect crystals. Rev. Mod. Phys. **36**, 681 (1964)
23. P. F. Fewster, *X-Ray Scattering from Semiconductors and Other Materials*, 3rd edn. (World Scientific, 2015)
24. C. Lichtensteiger, InteractiveXRDFit: a new tool to simulate and fit X-ray diffractograms of oxide thin films and heterostructures. J. Appl. Cryst. **51**, 1745 (2018)
25. S. Dolabella, A. Borzì, A. Dommann, A. Neels, Lattice strain and defects analysis in nanostructured semiconductor materials and devices by high-resolution X-ray diffraction: theoretical and practical aspects. Small Methods **6**, 2100932 (2022)
26. P.F. Fewster, X-ray analysis of thin films and multilayers. Rep. Prog. Phys. **59**, 1339 (1996)
27. J.E. Ayers, The measurement of threading dislocation densities in semiconductor crystals by X-ray diffraction. J. Cryst. Growth **135**, 71 (1994)
28. M.A. Moram, M.E. Vickers, X-ray diffraction of III-nitrides. Rep. Prog. Phys. **72**, 036502 (2009)
29. R. Uecker, B. Velickov, D. Klimm, R. Bertram, M. Bernhagen, M. Rabe, M. Albrecht, R. Fornari, D.G. Schlom, Properties of rare-earth scandate single crystals (Re = Nd−Dy). J. Cryst. Growth **310**, 2649 (2008)
30. A. Dhingra, D. Sando, P.S. Lu, Z.G. Marzouk, V. Nagarajan, P.A. Dowben, X-ray photoemission studies of $BiInO_3$: surface termination and effective Debye temperature. J. Appl. Phys. **130**, 025304 (2021)
31. P.F. Fewster, Reciprocal space mapping. Crit. Rev. Solid State Mater. Sci. **22**, 69 (1997)
32. U. Pietsch, V. Holý, T. Baumbach, *High-Resolution X-Ray Scattering from Thin Films to Lateral Nanostructures*, 2nd edn. (Springer, New York, 2004)
33. D. Kan, I. Takeuchi, Effect of substrate orientation on lattice relaxation of epitaxial $BiFeO_3$ thin films. J. Appl. Phys. **108**, 014104 (2010)
34. D. Sando, T. Young, R. Bulanadi, X. Cheng, Y. Zhou, M. Weyland, P. Munroe, V. Nagarajan, Designer defect stabilization of the super tetragonal phase in >70-nm-thick $BiFeO_3$ films on $LaAlO_3$ substrates. Jpn. J. Appl. Phys. **57**, 0902B2 (2018)
35. D. Sando, B. Xu, L. Bellaiche, V. Nagarajan, A multiferroic on the brink: uncovering the nuances of strain-induced transitions in $BiFeO_3$. Appl. Phys. Rev. **3**, 011106 (2016)
36. D.K. Bowen, B.K. Tanner, *High Resolution X-Ray Diffractometry and Topography* (CRC Press, London, 1998)
37. A. Authier, S. Lagomarsino, B.K. Tanner, *X-Ray and Neutron Dynamical Diffraction*, 1st edn. (Springer, New York, NY, 1996)
38. K. Inaba, X-ray thin-film measurement techniques. I. Overview. Rigaku J. **24**, 10 (2008)

Synchrotron Techniques to Map Ferroelectric Domain Structures Including Superlattices

Arnoud Everhardt and Youngjun Ahn

Abstract Ferroic domains, intrinsic to various ferroic materials, are pivotal in dynamic responses to external mechanical, electrical, and magnetic stimuli. These domains offer profound insights into fundamental material properties due to their intimate coupling with order parameters. Recent advancements in thin-film growth techniques have led to the discovery of novel domain structures within superlattice heterostructures, marking a significant evolution in the study of ferroic domains. Synchrotron X-ray diffraction, characterized by its high coherence and brilliance, has been instrumental in advancing our understanding of ferroic domains and their dynamic behavior. This technique enables the precise measurement of domain diffuse scattering, allowing for the determination of domain periodicity, coherence, orientation, and their dynamic responses under applied fields. In this chapter, we offer an introductory guide to the fundamental concepts of ferroic domains and superlattice heterostructures, alongside an overview of synchrotron X-ray diffraction techniques. We provide practical guidance on the measurement and analysis of X-ray diffraction and domain diffuse scattering data. Additionally, we present exemplary research studies, detailing the characterization of parameters relevant to ferroic domain structures, thereby illustrating the application of these techniques in contemporary research.

Keywords Ferroic domain · Superlattice heterostructure · Domain diffuse scattering · Superlattice reflection

A. Everhardt
LioniX International, Enschede, The Netherlands
e-mail: a.s.everhardt@lionix-int.com

Y. Ahn (✉)
Materials Science Division, Argonne National Laboratory, Lemont, IL, USA
e-mail: youngjun.ahn@anl.gov

D. Sando et al. (eds.), *X-Ray Scattering Techniques for Epitaxial Oxide Thin Films*,
https://doi.org/10.1007/978-981-96-5945-6_3

1 Introduction

Synchrotron (XRD) is an excellent experimental technique to determine not just crystal structures with atomic-scale resolution, but also more types of long-range periodic arrangements in crystals or crystalline thin films. An outstanding example is the study of ferroelectric domains where the state (orientation or magnitude) of the order parameter (polarization) changes with a spatial period across crystals. The ferroelectric domains directly contribute to electronic, optical, and dynamical properties of the material system. Thus, the investigation of this long-range configuration and dynamics can provide key aspects to understand important fundamental physics of ferroelectric systems. Synchrotron techniques with the highly coherent and brilliant X-ray radiation source allow systematic investigation by exploiting so-called domain diffuse scattering arising from periodic structural modulations coupled to the ferroelectric domains. The scope of this chapter is to introduce the synchrotron XRD techniques to study domain diffuse scattering to extract information on this type of periodic arrangement and to provide practical guidelines important for the synchrotron XRD experiments.

1.1 Ferroic Domains

Ferroic materials are classes of materials which typically describe the classes of ferromagnetic, ferroelectric and ferroelastic materials [1]. What all these materials have in common is that they have an order parameter [2] such as magnetization, polarization or strain, a vector which has (1) a magnitude and (2) a direction in every unit cell of the material.

Domain is defined as a region with the spatially uniform order parameter, which is naturally formed as the result of an energetic balance of competing components. Nature has a method to reduce the buildup of too large fields by forming the period arrangement of domains with different vectors of the order parameter. The smaller the periodic domain size, the lower the total energy cost since neighboring domains partially cancel out their respective fields. On the other hand, the formation of different domains costs energy by necessitating the formation of domain walls where a spatial gradient of the order parameter forms, which costs energy. From the balance between these two factors, an optimum domain width w can be determined following the Kittel's law

$$w = \beta\sqrt{d} \tag{1}$$

where d is the thickness of the material and β is a constant which depends on the material system [3, 4] and is related to the domain wall formation energy. In ferromagnetic systems, β is relatively large ranging between 10 and 100 $nm^{1/2}$ [4]. Due to the following reason: in ferromagnetic materials, the energy cost for the spatial

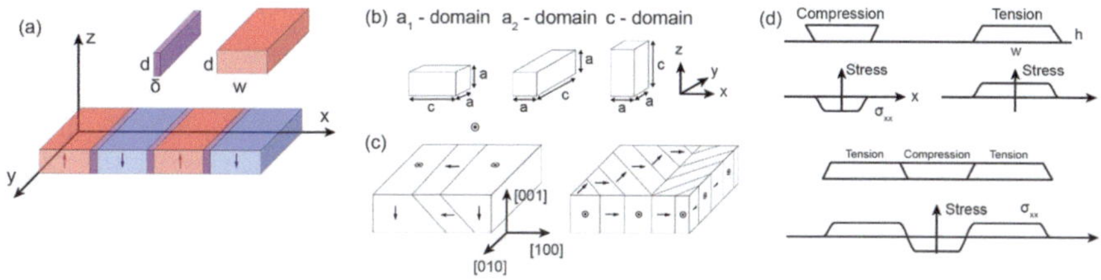

Fig. 1 **a** Typical geometry of 180° ferroelectric domains. The polarization in the domains alternates from up to down across a domain wall as indicated by red and blue arrows. **b** Types of tetragonal ferroelastic domains. **c** a/c- (left) and a_1/a_2-type (right) ferroelastic domain configurations. **d** In-plane stress profiles σ_{xx} of c-domain (compression) and a-domain (tension) in a/c-type domain and the energy minimization on the creation of a/c/a domain structures with alternating stress profiles across domain walls. Figure reproduced from Refs. [3, 4, 7]

gradient of magnetization is relatively moderate. Therefore, the magnetization across domain walls smoothly rotates from one direction to the other while the energy cost for the domain wall is relatively low by creating wide domain walls (> 1–10 nm). In contrast, ferroelectric systems have β values ranging between 1 and 5 $nm^{1/2}$ [4–6]. Polarization is strongly coupled to crystal symmetry and can therefore barely create a smooth rotation across the domain wall to connect the two polarizations. Therefore, the domain wall energy is relatively high, and thus the walls are narrow (< 1 nm). Ferroelectric films with thicknesses of 10–500 nm generally have domain sizes on the order of 5–100 nm. In single crystals of a millimeter thick, the domain sizes can scale up to 5 μm.

One of representative ferroelectric domain types is the 180° ferroelectric stripe domain, shown in Fig. 1a. This is a purely ferroelectric domain wall, where the polarization (order parameter) is alternating up and down across a domain wall. These domains form in ferroelectric single crystals and thin films with out-of-plane polarization in order to reduce the total electric field energy.

A hybrid domain type consisting of two order parameters is the ferroelectric-ferroelastic 90° domain. This is a typical domain pattern that forms when epitaxially growing ferroelectric thin films on a substrate [8] but it has also been shown to form in single crystals [9]. The driving force for the domain formation is the minimization of the elastic energy of the film on the substrate. Here we give the example of a tetragonal film where the c-unit cell parameter is larger than the a-parameter (a_1 and a_2). It forms, as shown in Fig. 1b–d, as so-called a/c or a_1/a_2 domains. The crucial part is that one of the domains gives positive (tensile) stress while the other domain gives negative (compressive) stress. Creating domains of alternating a and c (or a_1 and a_2) domains will result in mechanical boundary condition closer to an average zero stress to minimize the elastic energy (stress field) [7].

1.2 Superlattices

The boundary conditions of a ferroic material grown on a substrate can be exploited to form targeted domain patterns. Complex structures can be formed by artificially stacking different ferroic and non-ferroic materials alternatively in a so-called *superlattice* heterostructure consisting of a repeating stack of layers, as shown in Fig. 2.

Superlattices allow a knob to tune the internal electric fields by changing the layer thicknesses and material types to manipulate ferroelectric properties and long-range ferroelectric orderings [10, 11]. This can be combined with stress fields by growing epitaxial films on a suitable substrate for tensile or compressive stress to promote in-plane or out-of-plane polarization. Using these knobs, epitaxial $PbTiO_3$/$SrTiO_3$ superlattices have been a platform to realize theoretically predicted domain configurations ranging all the way from 180° stripe [12], flux-closure [13], polar vortex [14], to Skyrmion [15] domain patterns, as shown in Fig. 3.

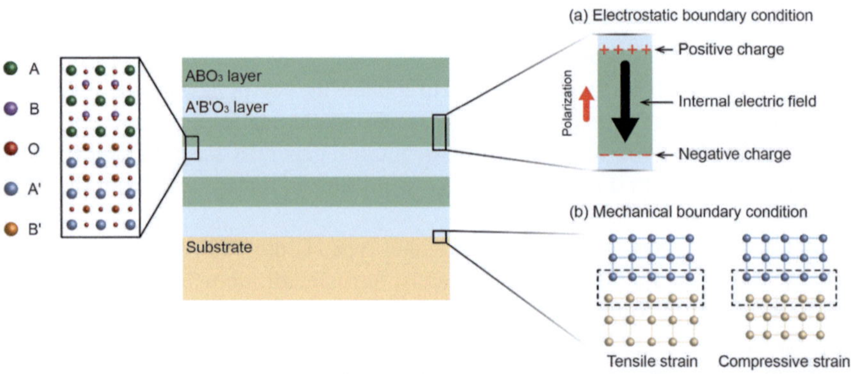

Fig. 2 A schematic of ferroelectric ABO_3/dielectric $A'B'O_3$ superlattice heterostructure epitaxially deposited on a substrate. **a** The polarization discontinuity at interfaces between ferroelectric and non-ferroelectric layers leads to internal electric fields opposite to the direction of polarization in ferroelectric layers. **b** Mechanical (elastic) boundary condition can be tuned by selecting suitable substrates for tensile or compressive stress on the superlattice layers

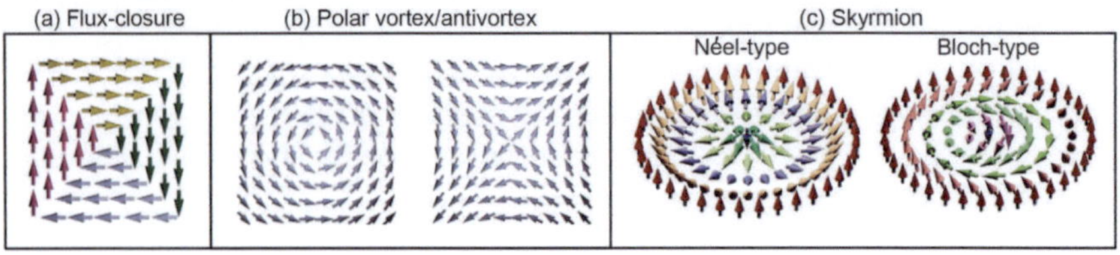

Fig. 3 Types of complex ferroelectric domain configurations realized in superlattice heterostructures: **a** flux-closure, **b** polar vortex/antivortex, and **c** Néel-type and Bloch-type Skyrmion domain configurations. The arrows represent directions of polarization. Figure adapted from Ref. [16]

Depending on types of ferroelectric domain configurations, not only can electrical and dynamical properties be altered [17–19] but also switching between different domain configurations can be driven by external stimuli [20–24], which makes the superlattice heterostructures an exciting platform in which researchers are able to discover novel phenomena.

1.3 Domain Diffuse X-Ray Scattering

Owing to periodic domain configurations ranging from nanometer to micrometer length scales, the investigation of structures and dynamics of domain configurations is challenging and limited to just a few experimental tools. Among these, XRD is specifically suited for periodic domain configurations. XRD technique provides an outstanding sensitivity to both atomistic structures on the Ångstrom-scale and longer-range structural orderings on the nanometer-scale. Ferroelectric/ferroelastic domain configurations with alternating polarization components exhibit spatial periods of structural modulations due to a strong polarization-structure coupling. Thus, XRD can be an excellent probe to investigate the domain configurations by studying so-called *domain diffuse scattering*.

Since the ferroelectric/ferroelastic domain patterns typically form along a crystallographic in-plane direction, the domain diffuse scattering arises around the main Bragg reflection with an in-plane distance as detailed in Sect. 3. The intensity and position of domain diffuse scattering in X-ray reciprocal space contribute to from many other physical components relevant to domain configurations in real space. The one contributing component is the periodic modulation of ions within the unit cell in each domain, and thus the polarization and its long-range configuration can be explored by examining the domain diffuse scattering. Several physical components such as domain wall configuration, coherent length of domains, and arrangement direction of domain patterns [25–27] also contribute to the domain diffuse scattering. The changes of ferroelectric domain patterns upon external stimuli have been investigated with the domain diffuse scattering, providing important insight into the domain configurations and their dynamics in ferroelectric/ferroelastic systems [20, 21, 28–31]. Therefore, studies of domain diffuse X-ray scattering can be an important pathway to understand physical phenomena relevant to the polarization and complex domain configuration and to discover novel physics.

1.4 Synchrotron X-Ray Diffraction

Domain diffuse scattering from nanometer-scale domain configurations is typically several orders of magnitude weaker than Bragg reflections from crystalline materials. Thus, a bright X-ray radiation source is crucial to properly measure domain diffuse scattering. Synchrotron radiation sources provide the most brilliant X-rays with intensities that exceed commercially available lab-based generators by a factor of 10^{10}. Furthermore, the brilliance of the synchrotron X-ray radiation source allows scientists to combine diverse techniques with XRD. For example, when combined with a proper choice of X-ray focusing optics, individual domains in nanoscale domain configurations can be selectively probed with focused X-ray beams [32]. Using coherent XRD technique combined with computational analysis techniques, domain configurations with a few nanometer periodicities can be spatially reconstructed [33]. In addition, sub-picosecond and sub-nanosecond X-ray pulses have allowed investigation of ultrafast dynamical behaviors of domain configurations [19, 32], utilizing pump-probe techniques. More recently, the effort to combine both temporally and spatially resolved techniques using synchrotron X-ray radiation source is in progress, which is expected to provide a characterization capability to probe temporal evolution of nanoscale heterogeneity [34].

More detailed discussion about synchrotron X-ray radiation source and facilities will be provided in Sect. 2 and the collection and interpretation of diffuse domain scattering will be further discussed in Sects. 2 and 3.

2 Experiment and Data Acquisition in Synchrotron Radiation Facility

2.1 Basic Principles of Synchrotron Radiation and Parameters of Synchrotron X-Ray Pulse

A synchrotron is a cyclic accelerator where a particle is accelerated close to the speed of light and travels along a closed-loop path. The particle is typically electrons and is injected into a storage ring after accelerating in linear accelerator and booster synchrotron. The storage ring is the closed-loop path where the electrons orbit with constant kinetic energy. The electromagnetic radiation from orbiting electrons is generated from the deflection of the electron by magnetic field from bending magnets. In order to enhance the total intensity of the radiation, an undulator is placed, followed by bending magnets. The undulator is an array of alternating dipole magnets along which the particles are forced to oscillate. At each oscillation, the electromagnetic fields are added in phase, resulting in the generation of coherent electromagnetic radiation source. Throughout this process, the synchrotron radiation sources are the most brilliant X-ray radiation sources with high coherence. The high brilliance and coherence make synchrotron X-ray radiation sources outstanding for research fields involving physics, chemistry, biology, and materials science. A selection of the representative synchrotron radiation facilities includes (i) the Advanced Photon Source in Argonne National Laboratory, USA, (ii) SPring-8, Japan, and (iii) the European Synchrotron Radiation Facility, France, which are classified as third-generation synchrotron radiation sources.

Here, specifications of the synchrotron X-ray radiation source, Advanced Photon Source in Argonne National Laboratory, IL, USA, are exemplified for the detailed description. The X-ray radiation source at Advanced Photon Source (APS) provides X-ray pulses with 90 ps pulse duration in 24-bunch mode in which 24 electron bunces travel in a storage ring with a temporal spacing of 153 ns. APS also provides 324-bunch mode for special purposes. In this mode, orbiting electron bunches are spaced with the spacing of 11.37 ns.

2.2 *Principles of X-Ray Diffraction and Experimental Geometry*

In this section, the details of experimental setup for synchrotron XRD measurements are discussed. XRD occurs when the elastically scattered X-rays from a periodic arrangement of electrons in materials interfere constructively. By virtue of Ångstrom-scale wavelength of X-rays, XRD can provide an atomic-scale probe of material structures. One exemplary schematic of the constructive interference condition for an atomic arrangement is shown in Fig. 4. Here, the X-ray scattering occurs from crystal planes composed of a periodic array of atoms. ω is an incident angle of X-rays, d is the spacing of two atomic layers, λ is wavelength of X-rays, and n is the diffraction order which is an integer. This condition of well-known $2d \sin \omega = n\lambda$ is called the Bragg condition where the constructive interference of scattered X-rays occurs. In this geometry, the (00l) reflections where (hkl) represents the Miller indexes can be measured because the scattering and constructive interference arise from a set of (00l) planes. Thus, by capturing scattered X-rays with detectors at a series of incident angles in this geometry, the d-spacing which is the distance between (00l) planes can be measured.

The experimental geometry for targeted (hkl) Bragg reflections can be visualized by considering the X-ray reciprocal space with wavevector coordinates $\boldsymbol{Q}_1$, $\boldsymbol{Q}_2$, and $\boldsymbol{Q}_3$[1] as shown in Fig. 5. Here, black dots represent reciprocal lattice points. X-rays with the wavevector $\vec{k}_{in}$ are incident on a sample at an incident angle ω and are scattered by a sample along the outgoing wavevector $\vec{k}_{out}$ with a scattering angle 2θ. When a wavevector $\vec{k}_{out} - \vec{k}_{in}$ from the (000) origin meets a targeted reciprocal lattice point by tunning the incident angle and scattering angle with respect to the sample, corresponding Bragg reflections can be measured by a detector. Since the distance between the reciprocal lattice points is inversely related to lattice spacings along [hkl] directions in real space, the lattice spacings can be measured by determining the position of Bragg reflections in the reciprocal space.

To satisfy the targeted Bragg conditions with non-zero h and k, a combination of independent rotational motions is often required. This can be fulfilled by employing a diffractometer. Figure 6 shows a schematic of 6-circle diffractometer with a capability

[1] In this chapter Q_1, Q_2, and Q_3 notations are interchangeably used with Q_x, Q_y, and Q_z to present results across the used references.

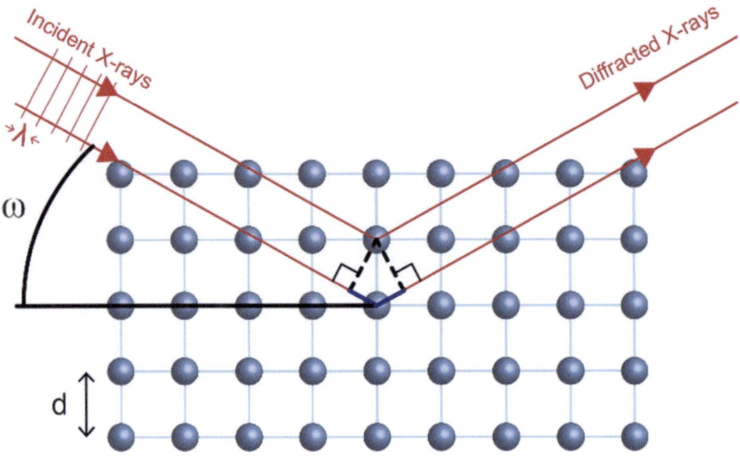

Fig. 4 A geometry of the Bragg condition: constructive interference of two X-ray beams incident at an angle ω with a wavelength of λ. Incident X-ray beams are elastically scattered from the atomic arrangement with a spacing of d. The path-difference between two X-rays beams is $2d \sin \omega$ which is displayed by blue lines

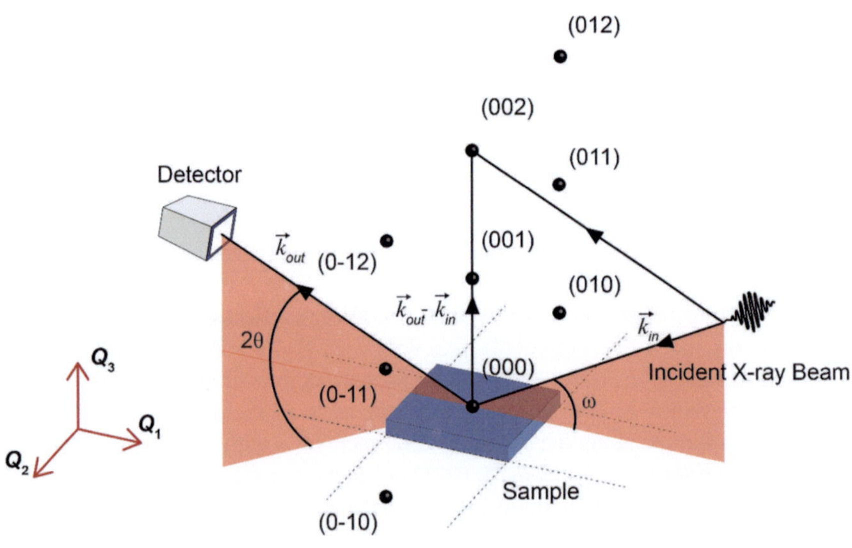

Fig. 5 Experimental geometry of X-ray diffraction for the (002) reflection of a crystal structure. Black dots presented here represent reciprocal lattice points. ω is the incident angle with respect to the sample surface, which is a half of the scattering angle 2θ under this condition. (hkl) represents the Miller indices of the corresponding crystal structure. In order to satisfy the condition for Bragg reflection with in-plane Miller index components, a combined rotation of the sample and the detector around the (00l) axis is required. Figure reproduced from [35]

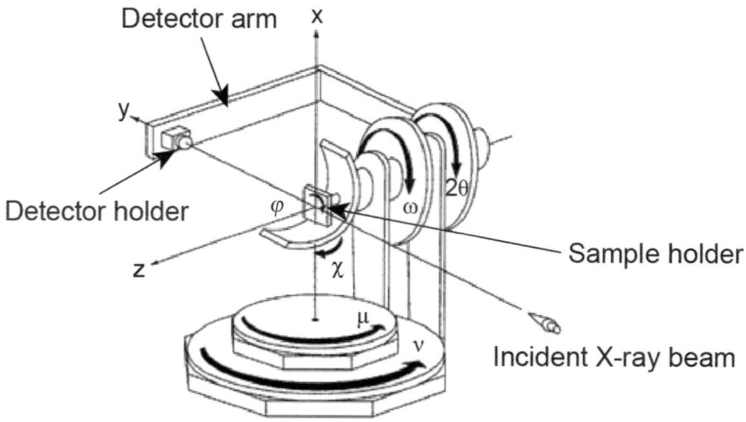

Fig. 6 A schematic of a 6-circle diffractometer with 2 detector-orienting and 4 sample-orienting capabilities. Figure reproduced from Ref. [36]

of four sample-orienting motions (φ, χ, ω, and μ) and two detector-orienting motions (ν and 2θ). At synchrotron facilities, the rotation of incident X-ray beams is not easily accessible. Therefore, the sample holder and the detector are rotated with respect to the incident X-ray beams. The geometry calculation for a (4 + 2) circle diffractometer is provided in Ref. [36].

For the purpose of measuring X-ray beams diffracted from the sample, two types of detectors are typically used: (i) avalanche photodiode and (ii) pixel-array area detector. The photodiode is often called a point detector which provides a single data point at each $\boldsymbol{Q}$ point with a precise reciprocal space resolution when combined with X-ray slits, featuring a fast time response and constant efficiency as well. The point detector is also capable of setting substantial gain, allowing the detection of extremely weak X-ray signal and the excellent signal-to-noise ratio. However, the point detector requires a long measurement time to obtain a full landscape of the three-dimensional reciprocal space. By comparison, the area detector can provide a two-dimensional cut of the reciprocal space, resulting in a significant reduction of acquisition time and additional flexibility for XRD geometries. However, the reciprocal space resolution provided by area detector is limited by a pixel size, which is expensive to reduce the size. The accuracy of the measurement can be also limited particularly when low-contrast sensitivity is required. One example of the area detector is the Pilatus 100K detector from Dectris. The Pilatus 100K detector is a two-dimensional pixel-array detector with a pixel size of 172 × 172 μm^2 and 94,965 pixels in total. The Pilatus 100K is electronically gateable, which is advantageous for time-resolved XRD measurements [37]. In the following section, an example of XRD measurements with an area detector is provided.

2.3 Procedures of X-Ray Diffraction Measurement with an Area Detector

This section provides an example of synchrotron XRD experiment of a (002) Bragg reflection using the Pilatus 100K detector so that readers can follow a given procedure at synchrotron facilities. First, procedures of the X-ray beam shaping and sample alignment with respect to the direct X-ray beams will be provided. Then, the assignment of angles to pixels on the area detector will be discussed. Finally, the reconstruction of the reciprocal space map with a series of detector images will be explained.

The size of direct X-ray beams is first manipulated by using a set of slits before a diffractometer. Sample size, X-ray focusing optics, and amplitude of targeted diffraction should be accounted for the pre-determination of the X-ray beam size. Typically, incident X-ray beams are cut to 50 μm width by horizontal and vertical slits. When a smaller X-ray spot size is necessary, the X-ray focusing optics such as Fresnel or refractive lenses can be used after the slits.

An alignment of a sample is required in order that the sample is placed at the rotation center of the diffractometer with respect to the direct X-ray beam, and an experimental geometry of $\omega = 0°$ can be determined. First, the sample surface needs to be placed at the half of the direct beam. This can be done by conducting direct X-ray beam half cut. Direct X-ray beam half cut is a measurement of the direct X-ray beam intensity as a function of Z as shown in Fig. 7. When the measured X-ray intensity is half of the maximum intensity of the direct X-ray beam, the sample is at the half of the direct beam. After placing the sample at the half of the direct beams, a tilt of the sample surface should be adjusted so that the sample surface along X and Y directions is perpendicular and parallel, respectively, to the incident X-ray beams. The tilt of the sample surface around the X axis can be confirmed by moving the sample along X direction and measuring the direct X-ray beam intensity. Given that the measured intensity before and after moving the sample is comparable, then the sample is well aligned. If the intensities at two different sample positions are significantly different, then the tilt must be compensated by rotating the angle χ. The tilt of the sample with respect to the Y axis can be checked by measuring the direct beams as a function of ω. At the angle at which the measured intensity is maximized, the sample surface is supposed to be parallel to the direct X-ray beams, which corresponds to $\omega = 0°$. After all these procedures, the half-cut measurement and the reiteration of the process described above are recommended to repeat.

The alignment procedure of the sample with respect to incident X-ray beam is generally performed using the direct X-ray beam. When the direct X-ray beams are used to align the sample, a set of attenuators must be inserted into the direct X-ray path to avoid degrading the sample quality and damaging the detector.

After the sample alignment, the assignment of angles and corresponding ***Q***s to pixels in the area detector is required to analyze the diffraction patterns. This procedure is unnecessary for a point detector because the point detector provides only a

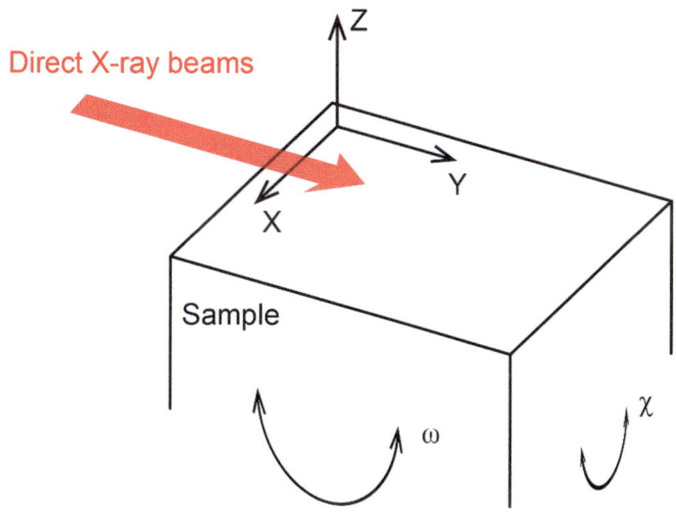

Fig. 7 Schematic of the rotation geometry of a sample with respect to direct X-ray beams. The direction of direct X-ray beams is perpendicular and parallel to the sample surface along X and Y directions, respectively. Tilting the sample surface around X and Y axes can be aligned by adjusting the angles ω and χ

single data point at one detector position. The area detector, however, provides two-dimensional diffraction profiles at one detector position. Thus, angle corresponding to each pixel in the two-dimensional array detector should be properly determined so that the diffraction patterns in the reciprocal space can be constructed for the analysis. The first step is a determination of the center pixel of the detector. The center pixel is a pixel at which the direct X-ray beam at $\omega = 2\theta = 0°$ is captured as shown in Fig. 8a. With respect to this center pixel, the angles of other pixels along vertical and horizontal axes are assigned. Angles $2\theta_{\text{Detector}}$ of each pixel along the vertical axis can be determined by an equation $2\theta_{\text{pixel}} = 2\theta + \tan^{-1}\frac{pV}{D}$, where V is number of pixels along the vertical axis away from the center pixel, p is a pixel size of 172 μm for the Pilatus 100K detector, and D is a distance between the sample and the detector. Corresponding α for each pixel can be obtained with an equation of $\alpha_{\text{pixel}} = \tan^{-1}\frac{pH}{D}$. D is typically obtained by physically measuring the distance between the sample and the detector or rotating the detector by $2\theta_0$ and then obtaining the shifted pixel numbers N from the center pixel with the direct X-ray beam so that $D = N/\tan 2\theta_0$. Note that D and p are important parameters when a high reciprocal space resolution is required for experiments. Either increasing D or reducing p can result in the higher reciprocal space resolution but placing the detector further from the sample is more easily approachable.

The center pixel at $\omega = 2\theta = 0°$ corresponds to the condition for the origin (000) as shown in Fig. 8c. Since the incoming incident X-ray beam is fixed, the sample and detector are rotated by ω_{002} and $2\theta_{002}$, respectively, as illustrated in Fig. 8d to meet the Bragg condition for the (002) reflection. The value of ω_{002} can be obtained

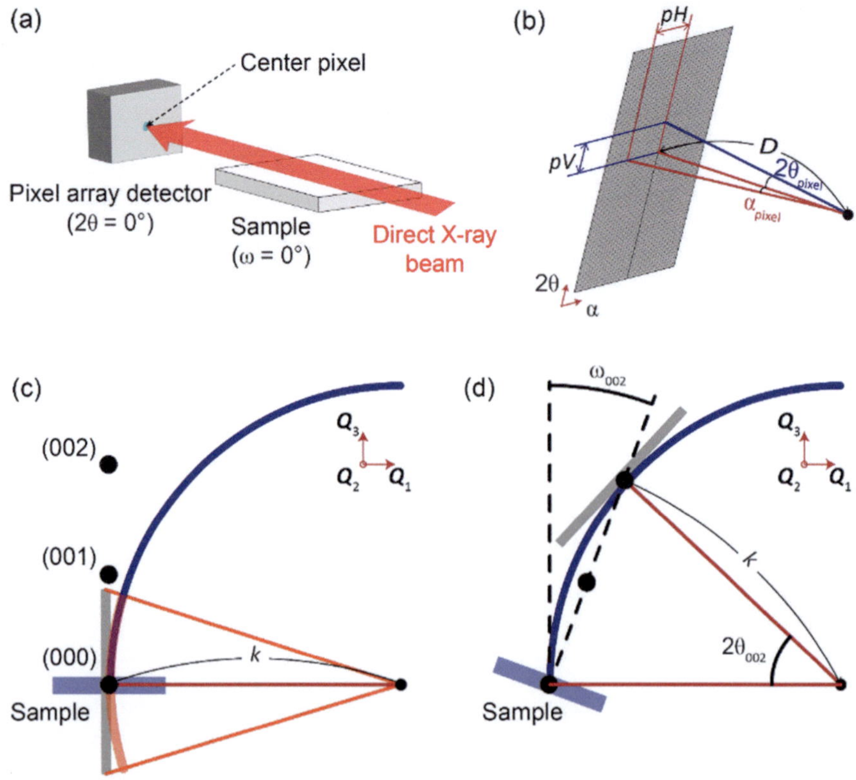

Fig. 8 **a** An experimental geometry with a pixel array detector at $2\theta = 0°$ and a sample at $\omega = 0°$. A pixel at which the direct X-ray beam is captured is defined as the center pixel. **b** A schematic of the way to assign corresponding angles $2\theta_{\text{pixel}}$ and α_{pixel} in the two-dimensional detector. **c** A schematic of the reciprocal space with reciprocal-space coordinates $\boldsymbol{Q}_1$, $\boldsymbol{Q}_2$, and $\boldsymbol{Q}_3$. The sample and area detector (a grey shaded box) are at $\omega = 0°$ and $\alpha = 0°$, corresponding to the geometry for the origin (000). A blue curve indicates a fraction of the Ewald sphere with a radius k. A slice of the reciprocal space within the area indicated by orange lines can be obtained at this detector position. **d** By rotating a sample and detector by ω_{002} and $2\theta_{002}$, respectively, the Bragg condition of the (002) reflection can be satisfied

with a known out-of-plane lattice parameter d by calculating the Bragg condition $2d \sin \omega_{002} = 2\lambda$. The (002) reflection should appear around the center pixel when $2\theta_{002} = 2\omega_{002}$. The diffraction pattern around the (002) reflection can be acquired by obtaining a series of detector images as a function of sample and detector angles with the condition of $2\theta = 2\omega$ in order to follow the (00l) axis.

After obtaining detector images around targeted reflections, three-dimensional reciprocal space along $\boldsymbol{Q}_1$, $\boldsymbol{Q}_2$, and $\boldsymbol{Q}_3$ axes can be reconstructed by stacking the detector images. The assigned angles for each pixel can be converted to corresponding $\boldsymbol{Q}_1$, $\boldsymbol{Q}_2$, and $\boldsymbol{Q}_3$ values and three-dimensional reciprocal space maps (RSM) can be reconstructed using publicly available RSMap3D program provided by the XSD-SDM division of APS [38]. A single two-dimensional area detector image can also be converted into an RSM by hand or a simple script by using the following set of formulas when $\varphi = \chi = 0°$:

$$Q_1 = k_0\{\cos(\alpha)\cos(\omega - 2\theta) - \cos(\omega)\}, \tag{2}$$

$$Q_2 = k_0 \sin(\alpha), \tag{3}$$

$$Q_3 = k_0[\sin(\omega)\{1 - \cos(2\theta)\cos(\omega)\} + \cos(\alpha)\cos(\omega)\sin(2\theta)], \tag{4}$$

where $k_0 = 2\pi/\lambda$.

By obtaining $\boldsymbol{Q}$-positions of the diffracted pattern or diffuse scattering from probed samples in two- or three-dimensional RSMs, atomic and mesoscale structures of probed samples can be obtained as follows. In RSMs, the axes are generally given in $\boldsymbol{Q}$. The units can be given as the inverse of a length scale, often Å^{-1}. This can be converted to real-space units by taking 2π and dividing that by the obtained $\boldsymbol{Q}$ value. Reporting the $\boldsymbol{Q}$ values in this way ensures that the units are independent of the X-ray wavelength: the angles in the XRD experiment depend on the wavelength λ, but they are in the end multiplied by $k_0 = 2\pi/\lambda$. So, this is the preferred notation for reporting but requires knowledge of the used wavelength. An alternative way is to report the $\boldsymbol{Q}$ values in relative lattice units (r.l.u.), which are the $\boldsymbol{Q}$ values divided by $2k_0 = 2\frac{2\pi}{\lambda}$. The interpretation of the two-dimensional image obtained by plotting $\boldsymbol{Q}_3$ together with $\boldsymbol{Q}_2$, and in a more general sense from the RSMap3D program, will be discussed in Sect. 3.

3 Analysis and Interpretation of Thin Film Domain Diffraction Data

The interpretation of domain diffuse scattering will be given with two examples of common ferroic domain types: 180 and 90° domains. These discussions will explain many of the features that can be seen on an RSM, and similar reasoning can be used to understand other domain structures as well.

3.1 180° Domain Diffraction Interpretation

A case study for the 180° domain wall from a superlattice heterostructure of $PbTiO_3$/$SrTiO_3$ [29] is displayed in Fig. 9a. $PbTiO_3$ is a ferroelectric material that, on substrates with in-plane lattice parameter roughly equal to the in-plane lattice parameter of $PbTiO_3$, would naturally form 180° domains with alternating up and down polarization. In contrast, $SrTiO_3$ is a dielectric material which does not naturally polarize. However, the material is highly susceptible to internal electric fields, and thus the resulting energetic balance from the formation of this superlattice heterostructure induces artificial polarization in this dielectric layer. The $SrTiO_3$ follows the polarization domain structure of the ferroelectric $PbTiO_3$ layer with the same direction but with a smaller magnitude than the $PbTiO_3$ layers.

The XRD RSM from this type of superlattice is presented in Fig. 9b, and it contains important structural cues to obtain physical information on the domain structure. The first features of interest are the diffraction peaks along the Q_z direction, the purely out-of-plane component. They all derive from the (00m) Bragg reflection, and indeed the middle data set (l = 0 and m = 2) in Fig. 9b is the (002) Bragg reflection originating from the 'total' superlattice film. This $PbTiO_3/SrTiO_3$ superlattice heterostructure is 100 nm thick and consists of repeating units of 8-unit cells of $PbTiO_3$ and 3-unit cells of $SrTiO_3$. The out-of-plane lattice parameter of the total film can be observed at 3.13 $Å^{-1}$, corresponding to a real-space distance of 2.007 Å, which corresponds

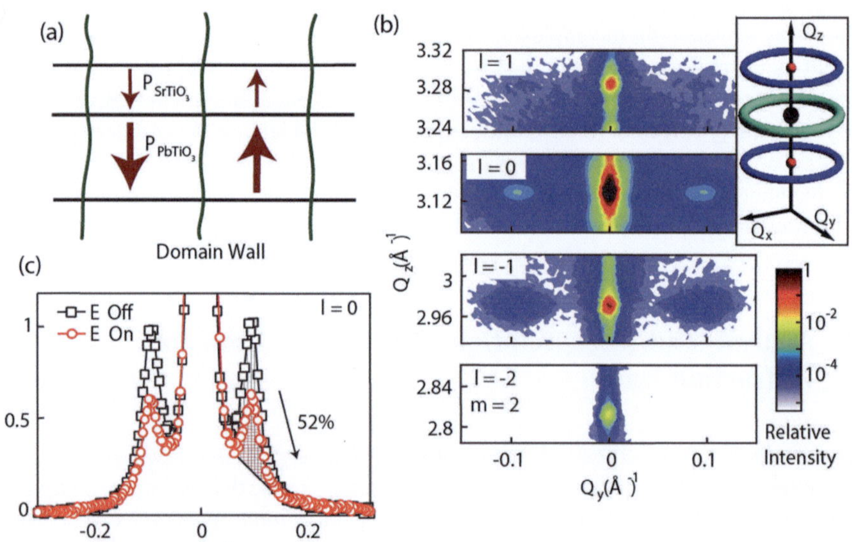

Fig. 9 **a** Schematic drawing of the domains resulting from the $PbTiO_3/SrTiO_3$ superlattice. **b** Reciprocal space maps around the (002) Bragg reflection (m = 2) presenting several superlattice peaks (l = − 2, − 1, 0, and 1). **c** Cross-section of domain diffuse scatterings along the Qy direction at the Qz at which the Bragg condition for the (002) reflection is satisfied. E represents an external electric field along the out-of-plane direction. l is the index of the superlattice reflections with l = 0 corresponding to the main (002) Bragg reflection. Positive and negative l represent superlattice reflections with higher and lower Qz positions with respect to the (002) Bragg reflection. Figure adapted from [29]

to the lattice parameter of 4.015 Å as the (002) reflection is presented here. On top of that, the superlattice structure with a repeating length of 11-unit cells creates a superlattice periodicity that results in additional peaks, denoted by integer number l, along the Q_z direction. The higher ($l = 1$) and lower ($l = -1$) superlattice reflections, at 3.28 and 2.97 $Å^{-1}$, can be seen, and they are observed with much lower intensity because a superlattice reflection is never as strong as the main reflection on which it is based. The distance between the main (002) reflection and the superlattice peaks is 0.15 $Å^{-1}$. This corresponds to a real-space distance of $2\pi/0.15\ Å^{-1} = 39$ Å. One repeating unit of the superlattice heterostructure has a total length of 44 Å, which is indeed close to this observed value.

The next feature in Fig. 9b is the diffuse scattering. The data at $l = 0$ and $m = 2$ shows two clear distinct peaks at non-zero $Q_y = \pm 0.1\ Å^{-1}$ around the (002) reflection at $Q_z = 3.13\ Å^{-1}$. The cross-sections are given in Fig. 9c. The diffuse scattering with non-zero in-plane wavevector components indicates that the scattering arises from the structural ordering arranging along in-plane direction in real space. The distance between the diffuse scattering and the Bragg reflection is ~ 0.1 $Å^{-1}$, and the value of the reciprocal of 0.1 $Å^{-1}$ multiplied by 2π is 6.3 nm, consistent with a period of the 180° domains in the film. The key feature is that these peaks are observed around the Bragg reflection with zero in-plane Q components, indicating that the structural modulation in the domains has out-of-plane components as consistent with the 180° domains. The position of the domain diffuse scattering is at the exact same out-of-plane values as the Bragg reflection due to the orientation of domain walls parallel to the out-of-plane direction. The domain diffuse scattering is also visible around the superlattice reflections with $l = -1$ and $l = 1$ because the domains form across both the entire film and repeating superlattice structure.

The fact that the domain diffuse scattering arises around the (00m) Bragg reflection gives more information. Since the existence of the domain diffuse scattering around the Bragg reflection with $h = 0$ and $k = 0$ and $m \neq 0$ indicates the periodic structural modulation along the out-of-plane direction as described above, the domains here have a long-range polarization configuration with alternating up and down polarization. Alternating polarization directions along the out-of-plane direction indicate that the ions move up in one domain and down in the other domain. It represents that the domain diffuse scattering presented here is sensitive to these types of structural modulations. From this discussion, it naturally follows that these 180° domains would not produce any domain diffuse scattering around (hk0) in-plane Bragg reflections without out-of-plane components. There is no in-plane structural modulation in this type of domain, and thus the domain diffuse scattering cannot exist around the Bragg reflections with $m = 0$. Alternatively, an a_1/a_2 domain structure with alternating in-plane polarizations and no out-of-plane components does not allow to produce domain diffuse scattering around (00m) Bragg reflections. Off-axis peaks such as a (103) Bragg reflection will give domain diffuse scattering for both domain types because both an in-plane and out-of-plane component is present in the (103) Bragg reflection.

An important implication of the fact that domain diffuse scattering arises from structural modulations in domains is that the intensity of the domain diffuse scattering is sensitively coupled to the polarization. Thus, the domain diffuse scattering provides a route to detect existence and modulation of the polarization configuration in ferroelectric thin films. For example, the domain diffuse scattering was used to examine ferroelectricity in thin films down to 3-unit cells with which electrical measurements of the polarization are unavailable due to a large leakage current in thin films [12]. An example of studies of externally modulated domain configuration is shown in Fig. 1c where an electric field is applied to the material and, as a consequence, the intensity of the domain diffuse scattering is reduced. Here, an electric field is an external force that modulates direction and magnitude of polarization in ferroelectrics. An applied electric field on ferroelectrics with the 180° domains changes the sign of the polarization in one of the domains while leaving the other intact. A small applied electric field partially changes the polarization, leading to the inhomogeneous polarization configurations in the domains. As a result, the spatially inhomogeneous structural modulations decrease the intensity of the domain diffuse scattering. If polarization of all domains was switched to one direction by a sufficiently strong electric field, the diffuse scattering should disappear as there is no structural periodic modulation along the out-of-plane direction.

The last feature from the data in Fig. 9b is in the three-dimensional drawing as shown in the inset which gives the Q_x, Q_y and Q_z directions. This drawing is made based on additional measurements which are not shown here but from which a three-dimensional RSM is constructed. The black dot is the Bragg reflection, and the red dots are the superlattice peaks. The key feature in this inset is the rings corresponding to the domain diffuse scattering around their respective Bragg or superlattice peaks. The ring shape of the domain diffuse scattering indicates that the direction of the in-plane scattering can take any value. The implication for the domain structure is that there is no preferred in-plane direction of the domains and domain walls, and they form in all directions. These domains are formed as circles or even completely irregular shapes such as a labyrinth with an average domain period of 63 Å along any in-plane directions.

3.2 90° Domain Diffraction Interpretation

The second case study is of a 90° a/c domain structure. A prototypical example is the ferroelastic a/c domain structure in a $PbTiO_3$ thin film grown on $DyScO_3$ (Fig. 10a). A feature of the domain structure in the $PbTiO_3$ on $DyScO_3$ is a portion of c-domain larger than a-domain, resulting from a smaller in-plane lattice mismatch of c-domain to the $DyScO_3$ lattice. Also, the a-domain is characterized by lattice tilts to accommodate the lattice mismatch at the domain walls. As a result, the polarization vector in the a-domains slightly tilts toward the out-of-plane direction.

The resulting XRD RSM around the (002) reflection is given in Fig. 10b with a butterfly-shaped diffuse scattering [39]. Here, the relative notation r.l.u. is used as

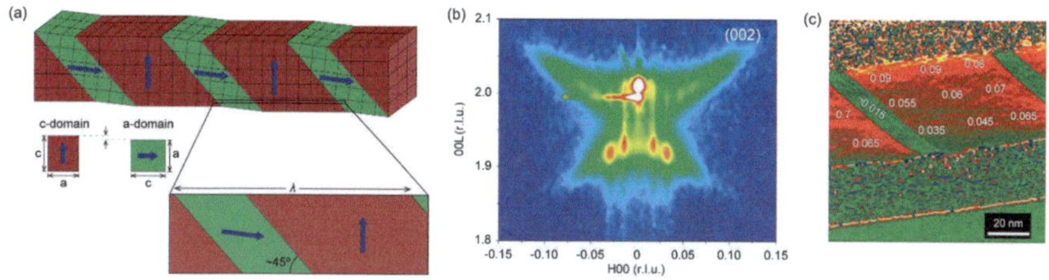

Fig. 10 **a** Schematic representation of ferroelastic a/c domain structures with a period of λ. The angle between domain wall and film/substrate interface is ~ 45°. The direction of polarization in a-domain presented by blue arrows is slightly rotated toward the out-of-plane direction to accommodate stress at the domain walls. **b** Diffuse scattering X-ray diffraction pattern of an a/c domain configuration around the (002) reflection of the $PbTiO_3$ thin film grown on $DyScO_3$ where $a_{PbTiO3} < a_{DyScO3} < c_{PbTiO3}$ corresponding to 3.9 Å < 3.95 Å < 4.15 Å. **c** Transmission electron microscopy cross-section of the a/c domain structure. The numbers represent the out-of-plane strain showing strain relaxation in the predominantly red-colored c-domains and green a-domains. Figure adapted from [39]

referenced to the $DyScO_3$ substrate, instead of $\boldsymbol{Q}$ coordinate. 1 r.l.u. corresponds to 1.59 $Å^{-1}$, which corresponds to $\frac{2\pi}{a_{DyScO3}}$ where the $DyScO_3$ lattice parameter is a_{DyScO3} = 3.95 Å. Along the out-of-plane wavevector direction at zero in-plane wavevectors, the substrate peak is observed at L = 2.0 r.l.u. indicating the second-order reflection.

From the two-dimensional RSM spanning along 00L and 00H directions, distributions of the several diffuse scattering can be observed. The diffuse scattering positioned lower than the substrate peak at L = 2 r.l.u. shows two pairs of strong peaks while the scattering at L > 2 r.l.u. shows relatively weak amplitudes. These are the diffuse scattering that arise from a/c domains: the peaks at L < 1.95 correspond to the c-domains with a longer out-of-plane lattice parameter and the peaks at L > 2 to the a-domains with a shorter lattice parameter than a_{DyScO3}. The diffuse scattering from a-domain can be observed around (002) reflection due to a small polarization component along the out-of-plane direction as described above. Weak and strong intensities of a-domain and c-domain diffuse scattering are attributed to two factors: (i) a larger portion of c-domain than that of a-domain in the a/c domain structure and (ii) much smaller out-of-plane polarization in a-domains. The total domain periodicity is determined to be 27 nm with 6 nm and 21 nm for the a-domain and the c-domain widths, respectively, by piezoelectric force microscopy (PFM) [40]. The in-plane periodicities have an in-plane spacing of 0.0138 r.l.u. for both the upper and lower branches of the butterfly shape. This corresponds to 3.95 Å/0.0138 r.l.u. = 29 nm, in good agreement with the domain sizes given above.

The orientation of the domain diffuse scattering lies diagonally in RSM. This diagonal diffusive distribution of the diffuse scattering is interpreted as a consequence of strain relaxation through the film thickness as well as domains and domain walls (Fig. 10c). The previous example of 180° domains in Sect. 3.1 shows domain walls perpendicular to the substrate. As a result, domain diffuse scattering forms purely along the in-plane directions with the same out-of-plane $\boldsymbol{Q}$ of the Bragg reflections.

In contrast, the a/c domains presented in this section have 45°-oriented domain walls with respect to the film/substrate interface. As a result, the domain diffuse scattering arises diagonally with 45° angle because both the domains and domain walls are the places from which the X-ray scattering occurs [31]. This provides a method to measure the domain wall angle with a high precision using high-resolution XRD.

4 Summary

To conclude, the RSMs from these case studies reveal that key structural information can be obtained from XRD data by understanding the origin of the different reflections and diffuse scattering. The analysis presented here can be readily translated to other systems with novel domain structures, promoting the understanding of the new structure with great detail. In addition, it is important to cross-reference other experimental techniques that are able to obtain structural data, such as high-resolution transmission electron microscopy (TEM) or atomic force microscopy (AFM)/PFM and to compare with theory and literature. XRD RSM analysis is a powerful tool, and this technique is exceptionally sensitive to tiny changes and small periodicities. As a cornerstone technique, the creative potential to extract plentiful detailed information from good quality materials is exceptional by studying the domain diffuse scattering.

Questions and Problems (Synchrotron Techniques to Map Ferroelectric Domain Structures Including Superlattices)

Q 3.1. Consider the discussion for Fig. 9.

i. Sketch how the normal coupled ($\theta - 2\theta$) X-ray diffraction would appear for this sample?
ii. How many superlattice peaks would you expect between the (001) and (002) Bragg reflections?
iii. The superlattice considered in the chapter has 8-unit cells of $PbTiO_3$ and 3-unit cells $SrTiO_3$. What is the average out-of-plane lattice parameter expected for a full c-axis oriented superlattice?
iv. Why do the authors double the observed real-space distance of 2.007 Å to extract the lattice parameter?
v. Plot how this average lattice parameter changes as function of $SrTiO_3$ content.
vi. Considering the same geometry and experimental set-up, plot how the distance between the main (002) reflection and the superlattice peaks changes as a function of $SrTiO_3$ content. Keep the number of $PbTiO_3$ cells constant at 8, and round the number of $SrTiO_3$ unit cells to the nearest integer. Compare your trend to available literature.

vii. Why do the authors come to the conclusion that the diffuse scattering in the Q_y directions are 180° domains? In the same way they state "an a_1/a_2 domain structure with alternating in-plane polarizations and no out-of-plane components does not allow to produce domain diffuse scattering around (00m) Bragg reflections". Sketch the a_1/a_2 domain structure and prove this.

Q 3.2. How does the diffuse scattering for 180° domains change as a function of applied field along the out-of-plane direction? Consider the conditions of zero electric field, coercive field, and saturation electric field.

Q 3.3. Consider a/c-type domain pattern aligned along the [010] crystallographic direction with a domain wall orientation angle of 45°. In what condition do you expect to find the corresponding domain diffuse scattering around a Bragg reflection positioned at Q_{Bragg}?

(a) $Q_{001} = Q_{Bragg}$; $Q_{h00} = 0$; $Q_{0k0} = 0$
(b) $Q_{001} = Q_{Bragg}$; $Q_{h00} \neq 0$; $Q_{0k0} = 0$
(c) $Q_{001} = Q_{Bragg}$; $Q_{h00} = 0$; $Q_{0k0} \neq 0$
(d) $Q_{001} = Q_{Bragg} \pm \alpha$; $Q_{h00} \neq 0$; $Q_{0k0} = 0$
(e) $Q_{001} = Q_{Bragg} \pm \alpha$; $Q_{h00} = 0$; $Q_{0k0} \neq 0$.

Q 3.4. Derive an equation for the orientation of the domain wall with respect to film/substrate interface, using positions of diffuse scattering and the (002) Bragg reflection in reciprocal space.

Q 3.5. Based on the arguments given to explain Fig. 10, how would you expect the diffraction pattern for a r_1/r_2 type in a rhombohedral $BiFeO_3$ crystal to look?

Q 3.6. Consider that you aim to measure the diffuse scattering for 180° domains with alternating out-of-plane polarizations in a $PbTiO_3$ film. In order to resolve the diffuse scattering around the (002) Bragg reflection, a distance between the Bragg reflection and the diffuse scattering is required to be at least ten pixels in a two-dimensional pixel array detector (Pilatus 100K). How far should the detector be positioned from the sample to attain the goal? Use the following parameters:

- Out-of-plane lattice parameter of $PbTiO_3$: 4.29 Å
- X-ray photon energy: 11 keV
- Periodicity of 180° domain pattern: 100 nm.

Q 3.7. At the Advanced Photon Source, one mode of operation is to use 24 bunches of electrons, each with a pulse duration of 90 ps with 153 ns spacing between the pulses. Using this information, calculate the total time that it takes for an electron to circle the storage ring.

Q 3.8. Compare and contrast the available spectral distribution of synchrotron radiation for at least two synchrotron sources in the world.

References

1. X. Hao, J. Zhai, L.B. Kong, Z. Xu, A comprehensive review on the progress of lead zirconate-based antiferroelectric materials. Prog. Mater Sci. **63**, 1 (2014)
2. K.G. Wilson, Renormalization group and critical phenomena. I. Renormalization group and the Kadanoff scaling picture. Phys. Rev. B **4**, 3174 (1971)
3. A.K. Tagantsev, L.E. Cross, J. Fousek, *Domains in Ferroic Crystals and Thin Films* (Springer, New York, 2010)
4. G. Catalan, J. Seidel, R. Ramesh, J.F. Scott, Domain wall nanoelectronics. Rev. Mod. Phys. **84**, 119 (2012)
5. G. Catalan, J.F. Scott, A. Schilling, J.M. Gregg, Wall thickness dependence of the scaling law for ferroic stripe domains. J. Phys. Condens. Matter **19**, 022201 (2007)
6. G. Catalan, I. Lukyanchuk, A. Schilling, J.M. Gregg, J.F. Scott, Effect of wall thickness on the ferroelastic domain size of $BaTiO_3$. J. Mater. Sci. **44**, 5307 (2009)
7. W. Pompe, X. Gong, Z. Suo, J.S. Speck, Elastic energy release due to domain formation in the strained epitaxy of ferroelectric and ferroelastic films. J. Appl. Phys. **74**, 6012 (1993)
8. D.G. Schlom, L.-Q. Chen, C.-B. Eom, K.M. Rabe, S.K. Streiffer, J.-M. Triscone, Strain tuning of ferroelectric thin films. Annu. Rev. Mater. Res. **37**, 589 (2007)
9. A. Schilling, R.M. Bowman, J.M. Gregg, G. Catalan, J.F. Scott, Ferroelectric domain periodicities in nanocolumns of single crystal barium titanate. Appl. Phys. Lett. **89**, 1 (2006)
10. M. Dawber, C. Lichtensteiger, M. Cantoni, M. Veithen, P. Ghosez, K. Johnston, K.M. Rabe, J.M. Triscone, Unusual behavior of the ferroelectric polarization in $PbTiO_3/SrTiO_3$ superlattices. Phys. Rev. Lett. **95**, 177601 (2005)
11. E. Bousquet, M. Dawber, N. Stucki, C. Lichtensteiger, P. Hermet, S. Gariglio, J.M. Triscone, P. Ghosez, Improper ferroelectricity in perovskite oxide artificial superlattices. Nature **452**, 732 (2008)
12. D.D. Fong, G.B. Stephenson, S.K. Streiffer, J.A. Eastman, O. Auciello, P.H. Fuoss, C. Thompson, Ferroelectricity in ultrathin perovskite films. Science **304**, 1650 (2004)
13. Y.L. Tang et al., Observation of a periodic array of flux-closure quadrants in strained ferroelectric $PbTiO_3$ films. Science **348**, 547 (2015)
14. A.K. Yadav et al., Observation of polar vortices in oxide superlattices. Nature **530**, 7589 (2016)
15. S. Das et al., Observation of room-temperature polar Skyrmions. Nature **568**, 7752 (2019)
16. S. Chen et al., Recent progress on topological structures in ferroic thin films and heterostructures. Adv. Mater. **33**, 2000857 (2021)
17. P. Zubko, J.C. Wojdeł, M. Hadjimichael, S. Fernandez-Pena, A. Sené, I. Luk'yanchuk, J.-M. Triscone, J. Íñiguez, Negative capacitance in multidomain ferroelectric superlattices. Nature **534**, 1 (2016)
18. A.K. Yadav et al., Spatially resolved steady-state negative capacitance. Nature **565**, 7740 (2019)
19. Q. Li et al., Subterahertz collective dynamics of polar vortices. Nature **592**, 7854 (2021)
20. Y. Ahn, J. Park, A. Pateras, M.B. Rich, Q. Zhang, P. Chen, M.H. Yusuf, H. Wen, M. Dawber, P.G. Evans, Photoinduced domain pattern transformation in ferroelectric-dielectric superlattices. Phys. Rev. Lett. **119**, 057601 (2017)
21. V.A. Stoica et al., Optical creation of a supercrystal with three-dimensional nanoscale periodicity. Nat. Mater. **18**, 4 (2019)
22. P. Chen et al., Atomic imaging of mechanically induced topological transition of ferroelectric vortices. Nat. Commun. **11** (2020)
23. X. Li et al., Atomic-scale observations of electrical and mechanical manipulation of topological polar flux closure. Proc. Natl. Acad. Sci. U.S.A. **117** (2020)
24. S. Das et al., Local negative permittivity and topological phase transition in polar Skyrmions. Nat. Mater. **20**, 2 (2021)
25. A. Boulle, I.C. Infante, N. Leme, Diffuse X-ray scattering from 180° ferroelectric stripe domains: polarization-induced strain, period disorder and wall roughness. Res. Pap. **845** (2016)

26. J. Park, J. Mangeri, Q. Zhang, M.H. Yusuf, A. Pateras, M. Dawber, M. Holt, O.G. Heinonen, S. Nakhmanson, P.G. Evans, Domain alignment within ferroelectric/dielectric $PbTiO_3/SrTiO_3$ superlattice nanostructures. Nanoscale **10** (2018)
27. M. Hadjimichael, E. Zatterin, S. Fernandez-Peña, S.J. Leake, P. Zubko, Domain wall orientations in ferroelectric superlattices probed with synchrotron X-ray diffraction. Phys. Rev. Lett. **120** (2018)
28. J.Y. Jo, P. Chen, R.J. Sichel, S.J. Callori, J. Sinsheimer, E.M. Dufresne, M. Dawber, P.G. Evans, Nanosecond dynamics of ferroelectric/dielectric superlattices. Phys. Rev. Lett. **107**, 055501 (2011)
29. P. Chen, M.P. Cosgriff, Q. Zhang, S.J. Callori, B.W. Adams, E.M. Dufresne, M. Dawber, P.G. Evans, Field-dependent domain distortion and interlayer polarization distribution in $PbTiO_3$/ $SrTiO_3$ superlattices. Phys. Rev. Lett. **110**, 047601 (2013)
30. H.J. Lee et al., Structural evidence for ultrafast polarization rotation in ferroelectric/dielectric superlattice nanodomains. Phys. Rev. X **11** (2021)
31. Y. Ahn et al., Dynamic tilting of ferroelectric domain walls caused by optically induced electronic screening. Phys. Rev. Lett. **127**, 097402 (2021)
32. H. Akamatsu et al., Light-activated gigahertz ferroelectric domain dynamics. Phys. Rev. Lett. **120**, 96101 (2018)
33. S.O. Hruszkewycz, M.J. Highland, M. Holt, D. Kim, C.M. Folkman, C. Thompson, A. Tripathi, G.B. Stephenson, S. Hong, P.H. Fuoss, Imaging local polarization in ferroelectric thin films by coherent X-ray Bragg projection ptychography. Phys. Rev. Lett. **110** (2013)
34. H. Wen, M.J. Cherukara, M. Holt, Time-resolved X-ray microscopy for materials science. Annu. Rev. Mater. Res. **49**, 389 (2019)
35. Y. Ahn, *Photoinduced Structural Phenomena in Ferroelectric Oxide Electronic Materials Investigated by Synchrotron X-Ray Diffraction* (The University of Wisconsin-Madison, 2020)
36. H. You, Angle calculations for a '4S+2D' six-circle diffractometer. J. Appl. Crystallogr. **32**, 614 (1999)
37. A. Miceli, Application of pixel array detectors at X-ray synchrotrons. J. Instrum. **4**, P03024 (2009)
38. *RSMap3D | Advanced Photon Source*, https://www.aps.anl.gov/Scientific-Software-Engineering-And-Data-Management/Software/Rsm
39. G. Catalan, A. Lubk, A.H.G. Vlooswijk, E. Snoeck, C. Magen, A. Janssens, G. Rispens, G. Rijnders, D.H.A. Blank, B. Noheda, Flexoelectric rotation of polarization in ferroelectric thin films. Nat. Mater. **10**, 963 (2011)
40. A.H.G. Vlooswijk, B. Noheda, G. Catalan, A. Janssens, B. Barcones, G. Rijnders, D.H.A. Blank, S. Venkatesan, B. Kooi, J.T.M. Hosson, Smallest 90° domains in epitaxial ferroelectric films. Appl. Phys. Lett. **91**, 112901 (2007)

Coherence X-Ray Techniques for Nanoscale Characterization of Complex Oxides

Roopali Kukreja

Abstract Coherent X-ray scattering techniques have emerged as a versatile probe to investigate material behavior at nanoscale in both equilibrium and non-equilibrium conditions. Unlike conventional X-ray imaging methods that rely on complex optical components, coherent scattering is fundamentally limited only by the wavelength of the incident X-rays. The advent of new diffraction-limited sources, coupled with fewer optical constraints, establishes coherent X-ray scattering as a versatile and powerful tool for probing nanoscale dynamical processes. In this chapter, we explore predominant coherent X-ray scattering techniques with a particular focus on X-ray photon correlation spectroscopy (XPCS). XPCS utilizes coherent diffraction patterns, known as 'speckle patterns' collected as a function of time, to analyze material fluctuations and dynamics. These speckle patterns serve as unique fingerprints of the sample in reciprocal space, providing critical insight into nanoscale processes. We begin by discussing fundamentals of coherence and requirements for coherent X-ray scattering experiments including XPCS, Coherent Diffraction Imaging (CDI) and Ptychography. We then outline the key experimental considerations for XPCS including optics, sample requirements, detection scheme and other practical aspects of conducting an XPCS experiment. We also describe approaches for data analysis and interpretation for XPCS datasets. We discuss both one-time and two-time correlation functions, factors influencing correlation functions, fitting techniques and relationship between correlation functions and the dynamic structure factor. Finally, we examine the application of XPCS in probing fluctuations in two distinct complex oxides: charge and orbital ordering fluctuations in magnetite near the metal-insulator (Verwey) transition and ferroelectric domain fluctuations in low-strain Barium Titanate thin films. These studies highlight the wealth of information that XPCS and coherent scattering techniques can provide, underscoring their potential to significantly advance the dynamical pathways in correlated oxides.

Keywords Coherent X-ray scattering · X-ray photon correlation spectroscopy · Correlation functions · Equilibrium fluctuations · Non-equilibrium dynamics

---

R. Kukreja (✉)
Department of Materials Science and Engineering, University of California Davis, Davis, CA, USA
e-mail: rkukreja@ucdavis.edu

D. Sando et al. (eds.), *X-Ray Scattering Techniques for Epitaxial Oxide Thin Films*,
https://doi.org/10.1007/978-981-96-5945-6_4

1 Introduction

Coherent X-ray scattering techniques are a powerful tool to investigate nanoscale behavior including dynamical fluctuations in correlated oxides. Compared to the typical X-ray scattering methods (Bragg diffraction, small angle X-ray scattering (SAXS) etc.), coherent X-ray scattering uses coherence of the X-rays to obtain nanoscale information. The scattering of a fully coherent X-ray beam from a sample gives rise to interference effects which can be measured in the far field. This phenomenon is akin to Young's double slit experiment where the wavefront emanating from the double slits undergoes alternating constructive and destructive interference pattern on the detector due to coherence of the incident beam. The key difference being that the various regions of the sample scatter instead of two slits, resulting in a complex scattering pattern. This interference pattern dubbed 'speckle pattern' collected in the reciprocal space, contains a wealth of information about both the static and dynamical behavior of the material. In this chapter, we will discuss the basics of coherent X-ray scattering with an emphasis on X-ray photon correlation spectroscopy (XPCS).

While XPCS is a relatively newer technique for the X-ray regime and complex oxides, it has been widely utilized in the optical regime to study fluctuations and dynamical processes in amorphous materials such as polymers, colloids, glasses etc. In the optical regime, a visible laser is used to scatter of the sample and an interference pattern is measured as a function of time. The evolution of the interference pattern as a function of time provides insights into the dynamical behavior of the material, and thus the technique was named Dynamical Light Scattering (DLS). DLS has been adapted to the X-ray regime due to the following advantages,

- Ability to probe material at nanometer length scales due to shorter wavelength of the X-rays compared to the visible regime
- Relative transparency of X-rays to most materials which allows for higher penetration depth and thus ability to 'see' internal structure
- Comparatively weaker interactions with material, which results in suppression of multiple-scattering effects that often dominate the behavior in the visible regime
- The ability to combine XPCS with resonant scattering or Bragg diffraction which allows access to electronic, magnetic or structural ordering in correlated oxides.

In spite of these advantages, the development of XPCS or other coherent X-ray techniques has been limited. The key reason being lack of availability of highly coherent X-ray sources compared to optical laser sources, complexity of the data analysis algorithms to retrieve information from the rich speckle pattern, and the stringent demands on the stability and data acquisition quality of the experimental setup. The advent of extremely bright synchrotron sources and X-ray free electron lasers (XFELs) which provide highly coherent X-ray beams have made coherent X-ray techniques feasible. Also, there has been a strong push for the development of diffraction limited storage rings (DLSRs) which will provide fully coherent X-ray beam. The crucial science case driving this 'coherent revolution' is capturing dynamical behavior at the limits of ultrasmall and ultrafast in a wide variety of materials. Overall, these advances have resulted in the development of different methods to exploit information provided by fully coherent X-rays to understand subtle structural or electronic effects at the nanoscale which can dictate material functionality.

1.1 Coherence and Coherent Sources

In this section, we discuss the key concepts used to describe coherence properties. Coherence of light is usually associated with laser-like properties, where an established phase relationship exists between the wave measured at different points. Coherence is usually classified in two types, spatial (or transverse) coherence and temporal (or longitudinal) coherence. Spatial coherence refers to the correlation in the plane transverse to the propagation direction while temporal coherence refers to the correlation between the phases of a light wave at different points along the propagation direction [1–3]. The spatial coherence length is dictated by the divergence of the beam in the propagating direction, while the temporal correlation is dictated by the bandwidth and wavelength of the beam.

Typical sources which we encounter everyday (lamps, candles, sun etc.), spontaneously emit photons in a chaotic manner, both spatially and temporally [4, 5]. Nonetheless, coherent light can be obtained from such spontaneous sources by filtering the emitted light both spatially and temporally. As shown in Fig. 1, spatial filtering can be performed by positioning the slits or pinholes along the beam path, i.e. creating a point source. For example, stars act as an ‘effective point source’ for any observer on the earth, due to very high distance and low angular divergence. Thus, the optical signal arriving at earth is nearly a plane wave making laser interferometry feasible even though a star is a spontaneous source. Temporal filtering can also be performed by extracting a small energy bandwidth, for example using monochromators which can select a particular wavelength. However, both spatial and temporal filtering results in the loss of intensity. Thus, highly brilliant synchrotron or XFEL sources are required to generate coherent and high intensity X-rays. Note that the degree of coherence may vary between sources and the experimental setup (i.e. set of filters used) at different synchrotrons and XFEL.

Degree of coherence for an X-ray or optical beam is quantified in terms of the coherent volume which is given by temporal coherence length and spatial coherence length. As mentioned above, spatial coherence length (ξ_t) defines coherence properties of any source in the plane transverse to the direction of propagation and is dependent on the divergence of the propagating radiation [1–3]. The spatial coherence length (ξ_t) can be derived by considering the Young’s double-slit arrangement as shown in Fig. 2 [6]. The interference pattern obtained in the far field from a point source and the double slits results in fully developed maxima and zero intensity minima resulting in a 100% contrast. However, most sources have some lateral dimension and are not ideal point sources. Interference patterns obtained from non-

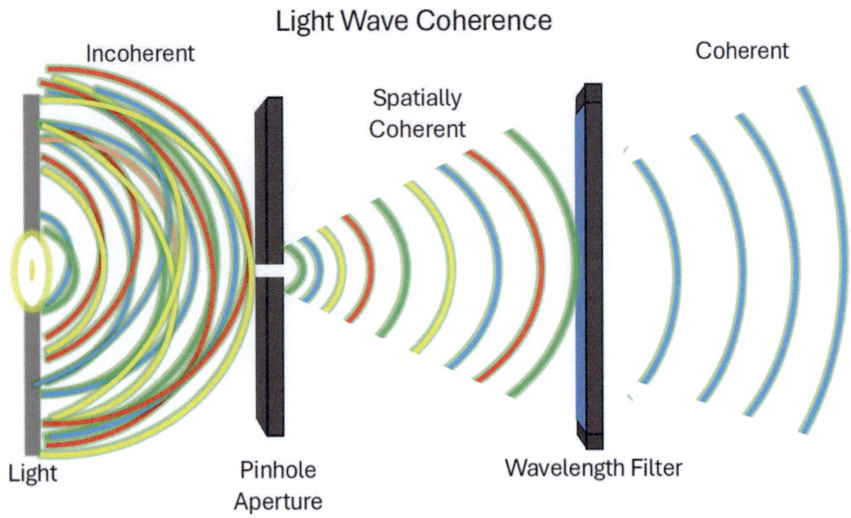

Fig. 1 Spatial and temporal filtering performed to achieve coherent beam [4, 5]

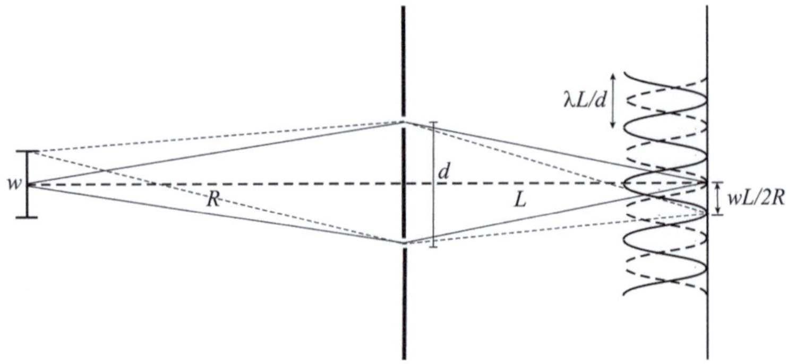

Fig. 2 Spatial coherence length can be obtained by considering Young's double slit experiment [6]

ideal real-life sources have lower visibility of the interface fringes. This is due to the fact that the interference pattern for each point of the non-ideal source emerges at slightly different diffraction angle. The superposition of these slightly shifted patterns will smear out the contrast lowering the fringe visibility. Using this method, the degree of coherence can be considered as the interference contrast measured in the far-field diffraction pattern.

Let us consider the interference pattern from a line source shown in Fig. 2. This line source can be considered as an array of infinitesimal point sources. The interference fringes obtained from the infinitesimal point source on the center of optical axis will be located at diffraction angles of $\alpha = \frac{m\lambda}{d}$ where m is an integer for maxima and m $= \pm$ 1/2, $\pm$ 3/2 … for the minima [6]. On the other hand, the fringes produced by the infinitesimal point source at the end of the line source will be shifted and appear at $\alpha = m\lambda/d + w/2R$. The distances are defined in Fig. 2. Small-angle approximations is used here. The maxima of point source at the end of the line will overlap with the minima from the point source at the center for a slit separation of $\lambda R/w$, and that distance is defined as the transverse coherence length,

$$\xi_t = \frac{\lambda R}{w} \tag{1}$$

Note that at a really far away distance (R goes to ∞) or for a point source ($w = 0$), the source will be fully coherent. For synchrotron sources with Gaussian intensity distribution the spatial coherence lengths are given by,

$$\xi_h = \frac{\lambda R}{(2\pi\sigma_h)} \qquad \xi_v = \frac{\lambda R}{(2\pi\sigma_v)} \tag{2}$$

for the horizontal (h) direction and the vertical (v) direction. Here σ_h and σ_v are horizontal and vertical widths of the source [7]. For synchrotron, a typical source size of $\sigma_v \sim 10 - 50\ \mu\text{m}$ and $\sigma_h \sim 100 - 500\ \mu\text{m}$ results in spatial correlation lengths of $\xi_v \sim 25 - 125\ \mu\text{m}$ and $\xi_h \sim 3 - 12\ \mu\text{m}$ respectively at a wavelength of $\lambda = 0.1$ nm and a distance of R = 40 m. In order to obtain fully coherent X-ray scattering the sample size should be smaller than the transverse coherence length. However, if the sample size is larger than the transverse coherence length, a part of the beam is selected for the experiment by placing a pinhole in front of the sample which results in close to fully coherent beam albeit with lower intensity. This practice is used at most coherent X-ray scattering beamlines to obtain close to fully coherent beam prior to the sample.

Temporal (or longitudinal) coherence length is defined along the direction of propagation and is dependent upon the wavelength [1–3]. It can be determined by considering the path lengths where two waves are completely out of phase as shown in Fig. 3 [6]. In this scenario, the first wave (λ) makes N oscillations in a given distance, and the second ($\lambda + \Delta\lambda$) wave should make ($N - 1/2$) oscillations. As the path covered by two waves is equal, $N\lambda = (N - 1/2)(\lambda + \Delta\lambda)$, which can be simplified and using substitution ($\xi_l = N\lambda$). Temporal coherence length is given by,

$$\xi_l = \frac{\lambda^2}{(2\Delta\lambda)} \tag{3}$$

This equation implies that the temporal coherence length is inversely proportional to the bandwidth of X-rays i.e. $\Delta\lambda/\lambda$. For a beamline with an X-ray wavelength of 0.1 nm and a typical bandwidth of $\Delta\lambda/\lambda = 10^{-4}$, this would result in $\xi_l \sim 380$ nm, which is much smaller than the spatial coherence length [2, 6]. Thus, in order to perform studies with a larger coherence volume, a monochromator is

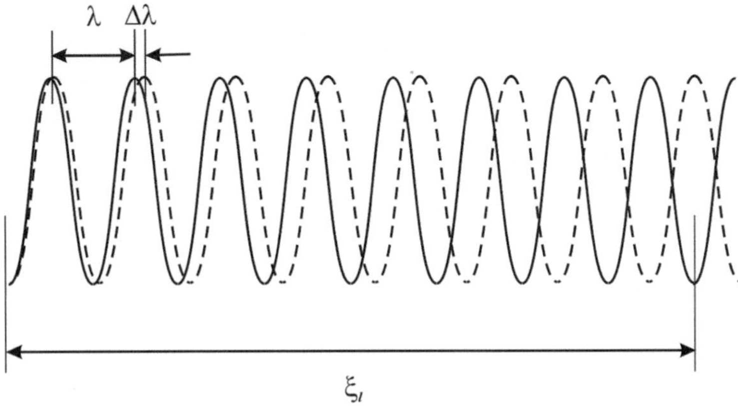

Fig. 3 Temporal coherence length can be obtained by considering propagation of two waves with wavelength λ and (λ + Δλ) [6]

used to select a particular wavelength and to increase the temporal coherence length. After monochromatization, the transverse coherent length of 1–2 μm is obtained for third generation undulator sources. Note that the two transverse and one longitudinal coherence lengths (ξ_h, ξ_v, ξ_l) are strongly dependent upon the X-ray optics, slits, etc. in the beam path. Thus, in practice, most experiments are performed with partially coherent radiation, and consequently, a quantitative understanding of interference or speckle pattern is critical [6, 8].

1.2 Speckle Pattern

As mentioned earlier, as coherent X-rays scatter from the sample, they generate an interference or 'speckle' pattern. Let us illustrate the idea of the speckle pattern with the simple case of visible laser light. When a laser beam (such as laser pointer beam) is incident on a projector screen or wall, it results in a nonuniform distribution of light, and appears speckled. In this case, the wavelength of laser light is too long to be scattered off air molecules, and instead the speckles are caused by roughness of the incident surface (wall, screen etc.). The laser light bounces off the incident spot at different heights or angles and interferes with itself, resulting in bright and dark regions. We can also visualize an experimental setup where the laser beam goes through a semitransparent material (tape, glass etc.), which is similar to the setup used for DLS. As the beam passes through the material, it acquires different path lengths, resulting in interference and the formation of the speckles pattern. This method in the visible regime was first used to investigate and understand Brownian motion.

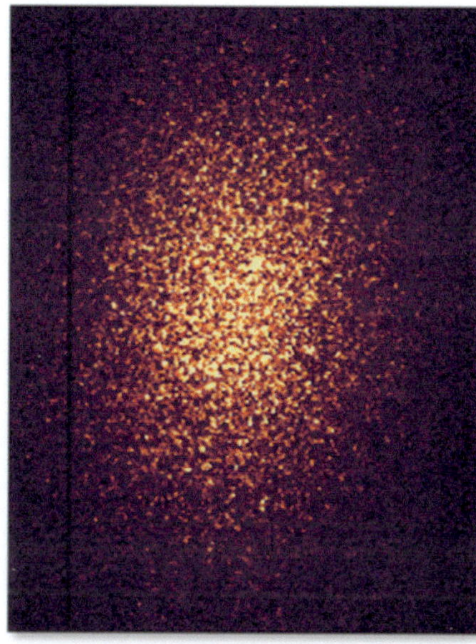

Fig. 4 Speckle pattern collected from a magnetite thin film (50 nm) measured at the (001/2) superlattice peak at the resonant Fe L-edge [9]

In the X-ray regime, the X-rays scatter from nanoscale disorder in the sample such as domains, domain walls, defect etc. Depending on the specific arrangement of the constituent atoms in the incident sample volume, X-rays may undergo constructive or destructive interfere after scattering to a certain angle. This results in the speckle pattern on the detector as shown in Fig. 4. It should be noted that the intensity at each pixel of the speckle pattern shown in Fig. 4 is the result of X-rays scattered from many different points in the sample. Path length of the X-rays originating from different points in the sample can be distinct and consequently, the resultant intensity at each pixel is due to sum of the X-rays with distinct phases. As the incident X-rays are coherent, the scattered X-rays with the distinct phase have a definite value at each point which results in constructive and destructive interference. On the other hand, if the incident X-rays are incoherent, there is no such phase relationship and the scattered intensity on each pixel will be an average value as shown in Fig. 5 [10]. Frankly, it is a good thing that conventional light sources which we encounter everyday are incoherent resulting in average scattered intensity, since 'seeing' speckle would be quite exasperating in everyday life.

From a fundamental perspective to study materials, the speckle pattern provides us with a unique way to achieve nanoscale dynamical behavior of materials. As the material undergoes changes (due to heating or cooling or other external stimuli), the speckle pattern will vary due to modification of interference pattern according to the new position of the scatterers within the sample. For example, the movement of domain walls under magnetic field will result in variation of speckle pattern. We can track these changes by collecting speckle pattern as a function of time. As the X-ray spot sizes are typically larger than atomic sizes, the illuminated volume contains

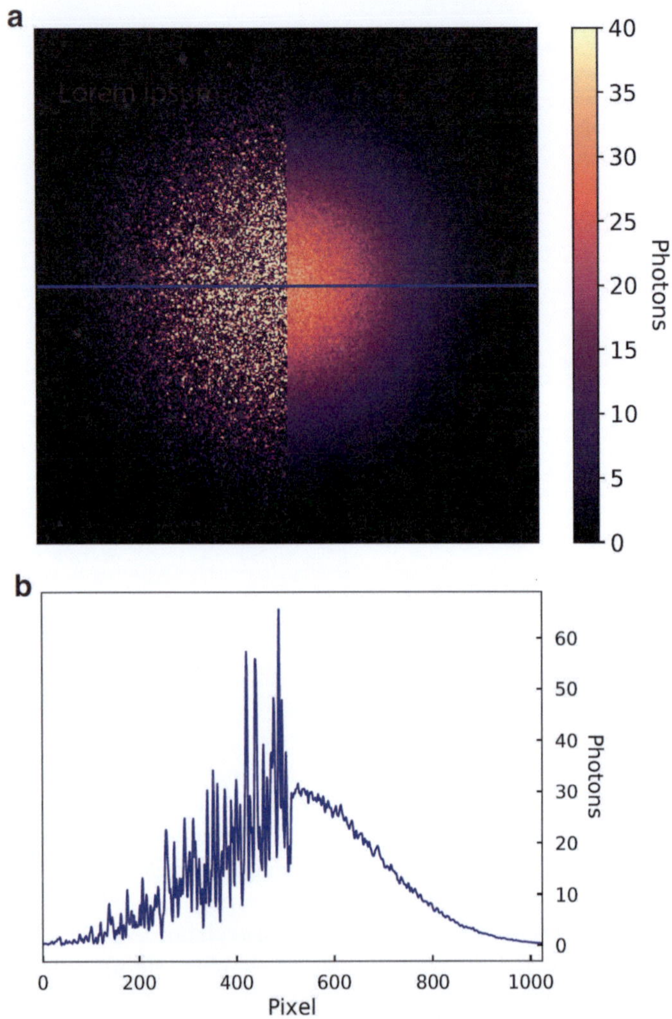

Fig. 5 Comparison of coherent versus incoherent scattering: **a** for a simulated 2D scattering pattern, and **b** a horizontal line cut through the 2D scattering pattern showing speckles (coherent scattering) on the left and smooth profile (incoherent scattering) on the right [10]

a huge number of scatterers, which can lead to very fast changes in the scattered intensity with time. These variations can be quantified by using correlation functions (see Sect. 3) and therefore, enable a glimpse of physical processes at nanoscale.

1.3 Coherent X-Ray Scattering Based Techniques

Before we discuss key coherent X-ray scattering based techniques, let us consider one of the main challenges involved in a scattering experiment. In a X-ray scattering experiment, where a sample is illuminated by an incident plane wave, the scattered wave in the far field (within the Born approximation) is the Fourier transform of the sample. The recorded intensity on the detector is proportional to $|F|^2$ where F is the structure factor dependent upon the electron density and its phase. However, as only intensity (or the magnitude of the Fourier transform pattern) is measured by the detector, the phase information is lost. In order to reconstruct the electron density of the object, the phase information must be known. This is the well-known phase problem in scattering, how do we retrieve the lost phase? The phase problem is treated different for crystalline and non-crystalline materials. For crystalline materials, Bragg diffraction resulting in sharp and bright Bragg peaks has been developed to reconstruct the crystal structure in the real space. However, the precise information regarding defects, domain walls, grain boundaries, local strain information is lost (as they are disordered in a crystalline material). For amorphous or non-crystalline materials, the diffraction pattern is faint and continuous. In this case, oversampling of the diffraction pattern has been suggested to reconstruct the lost phase information [11, 12] or pair distribution function is used to obtain information about nearest neighbours.

One of the key advantages of the coherent X-ray scattering techniques is the use of speckle pattern, where each pixel has definite relationship, to obtain the lost phase information. In XPCS, the speckles are tracked in reciprocal space to obtain real space dynamical behavior of the sample. In Coherent Diffraction Imaging (CDI), oversampling of diffraction pattern is used to reconstruct the phase and generate real space images using iterative phase retrieval algorithms. Depending upon the wavelengths of X-rays, different information can be extracted i.e. charge or magnetic order using X-rays tuned to resonant edges, or structural information using non-resonant hard X-rays or surface information using grazing incidence scattering. In the following, we describe each of these techniques briefly.

X-Ray Photon Correlation Spectroscopy (XPCS)

XPCS is a spatio-temporal technique where intensity fluctuations in the speckle pattern are used to measure the nanoscale dynamics. The evolution of intensity fluctuations in the speckle pattern can be directly related to fluctuations of the order parameter in real space. XPCS is ideally suited to study the relationship between kinetics and structure, and to measure fluctuations in equilibrium and non-equilibrium systems. One of the key advantage of the XPCS technique is the ability to simultaneously characterize macroscopic behavior by measuring a sample volume containing a large number of scatterers, while retaining the nanoscale spatial sensitivity defined by the size of the speckle pattern, which is typically of the order of 10's of nm. XPCS is thus ideally suited for studying nanoscale dynamics which cannot be accessed by visible light as well as most scanning probes which are limited to small lengthscales as shown in Fig. 6 [3].

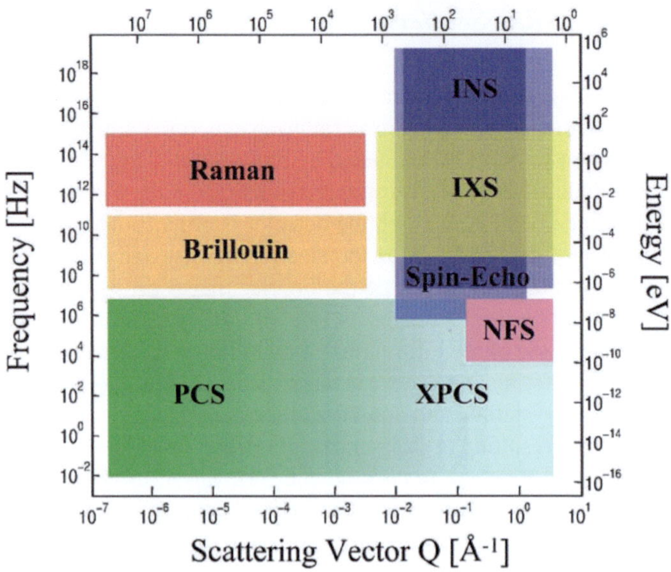

Fig. 6 XPCS compared to other spatio-temporal techniques [3]

XPCS has been utilized to study a variety of systems including polymers, non-Newtonian fluids, glasses and colloids [13–15], and fluctuations in antiferromagnetic metals [16, 17]. In addition to equilibrium structural fluctuations, this method has been applied to investigate defect migration during thin film growth [18, 19]. For complex oxides, XPCS has been utilized to investigate charge and orbital ordering fluctuations [9, 20, 21], superconducting systems [22, 23], and ferroelectric domain fluctuations [24, 25]. However, in most cases the window to study the fluctuations has been limited to time scales ranging from tens of milliseconds to thousands of seconds. This is mainly due to low coherent flux available or the limitation of readout times for the detector. Recent developments of the split and pulse technique, whereby X-ray pulse is split into two pulses, has been used at XFELs such as Linac Coherent Light Source to investigate fluctuations in magnetic order at femtosecond and picosecond timescales [26, 27]. In this chapter, we will focus on XPCS technique at synchrotron sources, but for readers interested in developments at XFEL sources, we point towards some of the emerging work [28–30].

Coherent Diffraction Imaging (CDI)

CDI is a lensless X-ray imaging technique which uses coherent X-ray scattering to achieve high spatial resolution without using high-quality lenses (such as zone plates) typical for X-ray microscopy technique [31]. Thus, the CDI technique a promising approach to overcome the resolution limit of direct X-ray microscopy techniques ($\sim$25 nm spatial resolution). CDI is based on the fact that a measured far-field speckle diffraction pattern contains information about large scattering vectors (q), which can

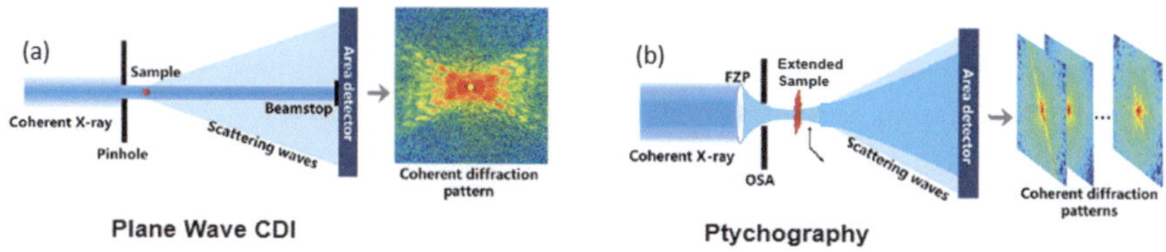

Fig. 7 CDI and ptychography: **a** CDI, where a plane wave illuminates the sample and the detector measures the oversampled diffraction pattern, and **b** ptychography, where focusing optics is used to scan the extended sample and diffraction pattern of partially overlapping regions of the sample are collected [31, 34]

be used to obtain nanoscale information limited only by the wavelength/diffraction-limit. Note that higher q in reciprocal space corresponds to smaller lengthscales in real space. As shown in Fig. 7a, the experimental setup uses high numerical aperture (by utilizing pinholes) to collect speckle pattern containing high q information [31]. Two key experimental approaches used to obtain phase information in CDI are, (a) oversampling of the diffraction pattern in reciprocal (or Fourier space), which implies the detector must have pixel size which is inversely proportional to the size of the object [12] and (b) the sample size is constrained i.e. the sample is illuminated by a known isolated illumination function [11, 32]. Iterative phase retrieval algorithms are then used to obtain the phase and reconstruct the images [11, 33].

CDI has been applied to a variety of nanomaterials to study structural behavior, for a review see Ref. [31]. Another version of CDI is Fourier transform holography (FTH), where a known reference scatterer (i.e. pinhole) is used with the sample. This results in a speckle pattern which contains interference of the known reference with the sample. Phase retrieval algorithms can then be used to obtain the initial phase and reconstruct the real space image. This method has been widely used to image magnetic domains in magnetic multilayered samples [35].

Ptychography

X-ray ptychographic microscopy combines the advantages of scanning X-ray microscopy method with CDI to achieve diffraction/wavelength limited spatial resolution [31, 36]. The key advantage of this technique is that the resolution is not limited by the fabrication challenges associated with X-ray lenses and optics (i.e. zone plate), nor by the requirement to isolate the sample. Ptychography, thus addresses challenges associated with X-ray microscopy techniques (specifically in the soft X-ray regime) such as scanning transmission X-ray microscopy techniques (STXM) or full-field transmission X-ray microscopy (TXM). Ptychography exploits the convolution in Fourier space, between the Fourier transform of a finite illumination function and the coherent speckle diffraction pattern of the object. In practice, this simply corresponds to measuring multiple diffraction patterns (Fig. 7) by raster scanning a finite (typically focused) coherent X-ray probe over the sample to collect a series of partially overlapping diffraction patterns [36]. This redundancy resulting from the oversampling condition provides a novel way to determine both the amplitude and phase of X-ray scattering and perform image reconstruction by using iterative reconstruction algorithms. The obtained real space complex-valued images provide a quantitative map of atomic displacements and enables characterization of strain, segregation, and dislocations within the sample.

Note that multiple methods fall under the overarching theme of CDI and ptychography and we point to some of the reviews articles for these methods [31, 36–38]. For the remaining chapter, we will focus on experimental design, data analysis and case studies for the XPCS technique.

2 Experimental Methods

As mentioned above, XPCS technique measures intensity fluctuations due to changes in speckle pattern using a 2D detector as a function of time (from ms to hours at synchrotron sources and from ps to ns at XFELs). Depending on the real space origin of the speckle pattern such as ordered domains, charge density wave (CDW) or pair distribution of a molecule, the intensity fluctuations can be related to specific dynamics inside the sample.

2.1 Experimental Geometry

Experimental geometries which have been typically used for XPCS are, (a) grazing incidence small angle X-ray scattering (GISAXS) in reflective manner, (b) Bragg geometry or wide angle X-ray scattering (WAXS) and (c) Small angle X-ray scattering (SAXS) in transmission geometry. Traditionally, XPCS studies have been performed in a GISAXS geometry as it provides access to smaller q-values or large length scales in real space, which is critical for materials such as glasses, polymers and liquids to study the long-range order. GISAXS geometry also provides high surface sensitivity which has been used to study Si and SiNx thin film growth [18, 19]. On the other hand, WAXS provides access to higher q-values or shorter atomic lengthscales. Specifically, in complex oxides it provides access to magnetic, structural, charge or orbital ordering by tuning to a specific Bragg or superlattice peak. Thus, it is a no surprise that WAXS-XPCS has recently become popular to study phase transitions in complex oxides. Figure 8 shows the experimental setup for the WAXS-XPCS [21]. SAXS in transmission geometry can be used to study thin film or transparent samples to access long range order or magnetic domain information. Note that transmission measurements require thinner samples (on SiNx membrane or free standing membranes) compared to WAXS-XPCS.

WAXS-XPCS studies for complex oxides have been performed in both resonant and non-resonant manner in hard and soft X-ray regime. Specifically, by tuning to resonant edges (such as L-edges for 3d transition metals, K-edge for oxygen) superlattice peaks (or forbidden reflections) due to charge, orbital or magnetic ordering

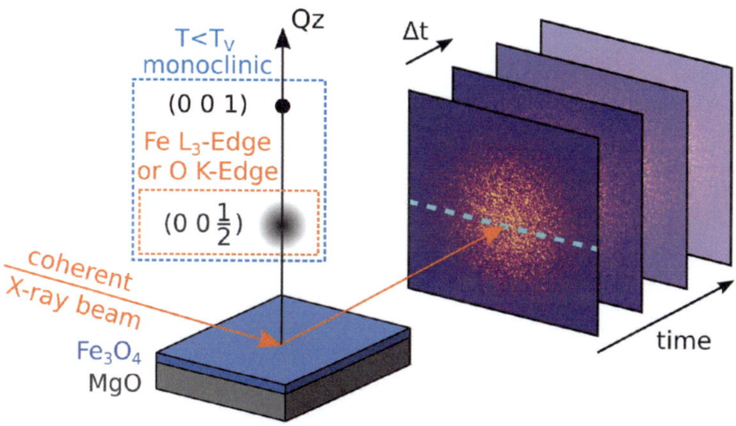

Fig. 8 XPCS experimental setup for the WAXS geometry for measurements performed [21] at CSX, NSLS-II

can be measured providing a unique way to access dynamical fluctuations in spin and electronic degrees of freedom across phase transition. Similarly, structural fluctuations can be measured by tuning to structural Bragg peaks usually in the hard X-ray regime. The choice of X-ray energy (resonant vs. non-resonant, hard X-ray vs. soft X-ray) also dictates which beamline or synchrotron to perform XPCS experiments. For example, XPCS measurements in hard X-ray regime can be performed at Sector-8-ID at Advanced Photon Source (APS), Coherent Hard X-ray Scattering beamline at National Synchrotron Lightsource II (NSLS-II), and Beamline ID10 (Troika) at European Synchrotron Radiation Facility (ESRF), at the time of writing this chapter. In soft X-ray regime, XPCS measurements can be performed at Coherent Soft X-ray Scattering (CSX) beamline at NSLSII, and COSMIC beamline at Advanced Light Source (ALS), at the time of writing this chapter. In the following, we provide a general overview of XPCS beamline setup including optics and X-ray detectors which were critical in improving the experimental stability. Detailed discussion on development in beamline optics and X-ray detectors suitable for XPCS studies can be found in Ref. [39] and references within.

2.2 *Incident Optics*

The incident optics of a beamline typically consists of an array of entry and exit slits, horizontal and vertical focusing mirrors and a monochromator to achieve a focused X-ray beam as shown in Fig. 9 [40]. Note that focusing distances for X-rays are much higher than optical wavelengths due to weaker X-ray-matter interactions. For example, at CHX at NSLS-II, the source to sample distance is 49 m. Typical X-ray beam size at a beamline is approximately 150 μm in vertical and 500 μm in horizontal

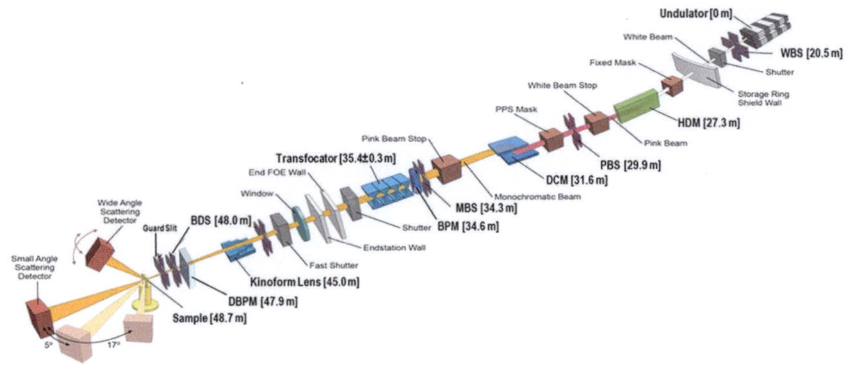

Fig. 9 Schematic of CHX beamline at NSLS-II [40]

direction. However, as discussed above, the spatial coherence lengths for X-rays are $\xi_v \sim 25 - 125$ μm and $\xi_h \sim 3 - 12$ μm. Thus, in order to achieve fully coherent X-rays for XPCS studies, a small part of intensity is selected using slits or pinholes to achieve X-ray beam size of $8 - 10$ μm, mainly limited by the horizontal spatial coherence length. However, utilizing slits or pinholes results in lowering of X-ray intensity and discarding a portion of X-ray beam with high coherence (i.e. in vertical direction) [39]. Vertical focusing of the X-ray beam can be used to compensate for this issue. A variety of high-quality focusing elements including refractive lenses, phase zone plates, and novel mirror designs have been developed to achieve submicron focused beam size (< 10 μm) [39].

In order to further increase the coherence, pinholes (varying from 3 to 15 μm) are used right before the sample for XPCS experiments. The size of this pinhole needs to be decided based on the requirements of the experiment, keeping in mind the trade-off between coherent volume and X-ray coherent flux. For photon hungry experiments, where the sample is either a weak X-ray scatterer or the signal to noise ratio is low, larger pinholes are used. Note that even for the smallest pinhole size the beam is still partially coherent as the longitudinal (temporal) coherence length is typically around 1 μm, dictated by the wavelength and bandwidth (see Eq. 3).

Thus, the second aim of focusing optics is energy selection which can be performed using filters, crystals and multilayers [39]. From the synchrotron undulator radiation, absorbing filters or mirrors are used to select a particular energy with a bandwidth ($\frac{\Delta\lambda}{\lambda} \approx 1.3\%$). However, for XPCS experiments, much stringent bandwidth is required and thus Bragg diffraction from perfect crystals is used to further narrow down the bandwidth i.e. Ge(111) Bragg reflection with bandwidth of $\frac{\Delta\lambda}{\lambda} \sim 3 \times 10^{-4}$. In order to further improve the temporal coherence length, Bragg scattering from multilayer mirrors can be used in the hard X-ray regime [39]. In the soft X-ray regime, monochromators with multilayer gratings are used as Bragg scattering cannot be used [41]. Figure 10 shows experimental schematic of the beamline at CSX beamline at NSLS-II [23]. However, use of monochromators results in lower intensity

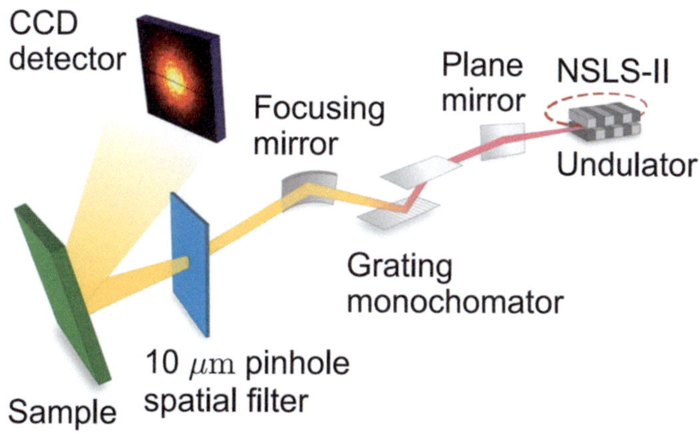

Fig. 10 Schematic of CSX beamline at NSLS-II [23]

making XPCS experiments more challenging, in addition to stringent requirements of ultra high vacuum due to higher absorption cross-section of soft X-rays.

2.3 Detection of Speckle Pattern

Weak signal levels in XPCS experiments place stringent requirements for the detectors including the ability to record small numbers of scattered photons with high fidelity. Furthermore, relatively high resolution and higher frame rate is required to resolve X-ray speckle pattern over a wider dynamic range for studying fluctuations and material dynamics. In fact, successful implementation of XPCS technique is strongly linked to the advancement in detector technology. While a point detector (such as PIN diode or scintillation counter) can be used to obtain intensity information with extremely fast readout time, it is limited to a single point (single q-value) and thus cannot capture the rich information embedded in a speckle pattern and also suffers from noise issues [39]. 2D detectors such as charge coupled devices (CCD) have been widely used as they provide most averaging over various scattering angles with a reasonable detection quantum efficiency in the 0.1–12 keV energy range. However, they have slower read out times. Currently, the best commercial XPCS area detectors are pixel array detectors (PADs) such as EIGER detector with 75 μm pixels and narrow-dynamic-range 0.5-megapixel frame rates of up to 9000 frames per second (fps), and LAMBDA detector with 55 μm pixels and anticipated narrow-dynamic-range 0.75-megapixel frame rates of up to 24,000 fps [42].

Another component that is critical for XPCS experiment is the placement of the detector. For XPCS measurements, typically the sample-to-detector distance is chosen such that the speckle size ($d_{speckle}$) is roughly the size of a pixel on the detector. If an individual speckle is larger than the pixel size, then the maximum intensity due to the constructive interference is spread over a larger area and the autocorrelation functions (see Sect. 3) are not able to completely track the speckle dynamics. If the speckle size is smaller than the pixel size, the resolution is lost. In both cases, the contrast of the measurement (brightest pixel to darkest pixel) is reduced. On the other hand for CDI, the speckle is typically oversampled i.e. speckle size is higher than the pixel size to recreate phase information [12]. The mean speckle size is given by following equation,

$$d_{speckle} = \frac{\lambda L}{\sigma} \tag{4}$$

where $d_{speckle}$ is the speckle diameter, λ is the wavelength of the X-ray beam, L is the sample-to-detector distance, and σ is the beam diameter. It should be noted that small speckles come from large regions, while large speckles come from small regions, and thus the smallest pixel measured is the largest lengthscale of the sample i.e. the full exposed/incident area. In an XPCS experiment, the sample-to-detector distance is typically 2 m and the aperture of the detector is 20 μm. This gives divergence of 10 μrad which implies that beam size must be in the 10 μm range for a 0.16 nm wavelength to maintain X-ray coherence on the detection side [39]. This is also one of the main reasons for choosing beam diameter in 5–10 μm range for XPCS experiments, so that coherence is also maintained on the detection side. Note that if X-ray beam is only partially coherent the visibility of the speckle decreases. A discussion of the statistical properties of the intensities of speckles and their visibility can be found in Ref. [3], which shows that the speckle contrast goes to zero as the coherence volume decreases relative to the scattering volume. Influence of partial coherence on XPCS analysis is also discussed in Ref. [8].

2.4 Other Experimental Considerations

In this section, we discuss some additional details to consider while conducting XPCS experiments.

Exposure Time and Measurement Window

Exposure time and measurement window is strongly dependent on both beamline and sample considerations. The exposure time for the detector (can vary from 0.1 to 5 ms) depends on the averaging needed to obtain good signal to noise ratio (SNR) from a given sample, and can be slower than the detector frame rate (i.e. 0.01 ms). Exposure time is strongly dependent on scattering from the sample for a given scattering vector in reciprocal space, and can limit the timescales of fast dynamics captured in an XPCS measurement. On the other hand, the longest timescales which can be measured are dictated by beamline stability which is typically around 3–4 h and thus limits the measurement window. Beamline stability is characterized using a reference static sample for the temperature range desired. Note that in order to accurately estimate characteristic decay constant for autocorrelation functions as discussed in Sect. 3, a measurement window of 3–5 times the decay constant is needed. Recently, detectors down to 0.1 μs have been commissioned at APS Sector-8. Both coherent flux and detection timescales are expected to increase by 10–100 times with the development of higher coherent next generation storage rings such as APS-U and ALS-U.

Coherent Volume and Diffraction Volume

In case of partially coherent beams, it is critical to optimize the speckle contrast as it dictates the SNR of the measurement. As mentioned above, X-ray beam spot size determined by the pinhole is one of the key consideration. The spot size also dictates the diffraction volume. While large volume usually implies higher SNR, larger diffraction volume (than coherence volume) reputism lower speckle contrast, which can reduce SNR. On the other hand, smaller diffraction volume (than coherence volume) could reduce SNR as it lowers macroscopic averaging of the sample behavior in the correlation function calculations. Thus, the key points which should be considered are [43], (i) the diffraction volume should be comparable in size to the coherence volume, (ii) diffraction pattern should be broad enough to capture lots of speckles in order to calculate correlation functions. For example, charge or orbital scattering peaks (superlattice peaks) of films are usually broader allowing for more speckles. The Bragg peak of a substrate is narrow and limited to few pixels resulling in less speckles. See Ref. [8] for further discussion on the limiting case of few speckles, (iii) intensity counts for the diffraction pattern per correlation time should be sufficient (a good rule of thumb is at least one count per speckle), (iv) sufficient number of correlation times should be measured to achieve good statistics from the sample.

Sample Details

For complex oxides, crystal orientation can strongly dictate the dynamics being studied, and thus detailed preliminary characterization via X-ray diffraction, such as Laue is needed to know the crystal orientation apriori. The typical sizes which are used for crystals and thin films for XPCS experiment are within range of 3–10 mm by 3–10 mm. For film thickness, while films of < 10 nm can be measured, the scattering is considerably weak, so thin film thickness > 25 nm is recommended. Other factors currently limiting the study of complex oxides using XPCS is achieving temperatures lower than 30–50 K and limited access to reciprocal space making it challenging to study in-plane or other asymmetric Bragg or superlattice peaks.

Beam-Induced Sample Damage

The measurement window for the sample can also be dictated by X-ray beam-driven sample degradation as the beam is focused on a 3–10 μm spot on the sample for 1–3 h. In order to track sample degradation, time-dependence of the average scattered intensity can be monitored (after normalization to incoming intensity). A systematic decrease in intensity of 7–10% have been typically observed for glass samples [44]. For polymer glasses or biomaterials, this degradation can be much worse and has been shown to induce radiolysis (beam induced bond breaking), crystallization and dynamics at cryogenic temperatures [45]. While for complex oxides radiation damage has not been observed so far, it possible that different sample parameters or the addition of organic molecules (i.e. hybrid perovskites) can incur beam-induced damage. In order to minimize sample degradation, the sample position can be adjusted to illuminate a previously unexposed spot on the sample or renewing the sample. Incoming beam intensity can also be lowered as sample damage is correlated with incoming X-ray intensity. Note that beam-induced sample damage can also initiate dynamics in the sample, making it challenging to differentiate inherent material dynamics from beam-induced dynamics.

2.5 Running a XPCS Experiment at the Synchrotron

Once all the above points have been considered for a given sample, we are now ready to perform XPCS experiments as enumerated below.

1. **Sample Alignment**: Alignment depends upon crystallographic axis and tuning to the Bragg peak or grazing incidence (SAXS, GISAXS, WAXS geometry) to access the required scattering vector in the reciprocal space. Note that if multiple scattering peaks are captured on a 2D detector (i.e. domain scattering peak for ferroelectric thin film [25]) while cutting through the center of one domain scattering peak, an off-center slice for other Bragg or scattering peaks might be captured that can result in different timescales. In that case, careful understanding of the reciprocal space or measurements at the center of multiple peaks is needed before interpreting the speckle dynamics.
2. **Stability of the Beamline**: Reference scans at room temperature and in the desired temperature range are critical for determining the stability of the beamline. Ideally, these would have been performed prior to the beamtime on a static reference sample such as aerogel. The static reference sample is used to establish the maximum contrast from the experimental setup and beam characteristics, as well as to measure the stability window over which the experimental setup shows minimum drifts or fluctuations. In addition to the reference, it is critical to collect XPCS scan for the desired sample in a temperature range where no fluctuations are expected, confirming the stability prior to the start of the measurement. However, note that if the autocorrelation curve (discussed in Sect. 3) is completely flat, that implies that either the sample is completely static or that the dynamics are too fast to resolve for the used detector settings. These two cases can be distinguished based on the value of autocorrelation, if value is closer to 1 then it implies that dynamics are too fast to capture, and if it is closer to the reference sample for that temperature, it implies the sample is static.
3. **Equilibrium Measurements**: XPCS scans are usually collected for a range of temperatures across a phase transition or as a function of external stimuli (stress, fields, etc.). In order to measure sample dynamics as a function of temperature, equilibration at each temperature is important. This is typically accomplished

by allowing 30–45 min for the sample and heating stage to achieve thermal equilibrium before starting the XPCS scan. The height and theta alignment to ensure sample is aligned should be performed for each temperature.

4. **XPCS Scans**: The overall measurement time for an XPCS scan as mentioned above is dependent upon the stability of the beamline and exposure time used for the detector, which in turn is determined based on sample dynamics. Typically, a 30 min–1 h scan is performed for a given exposure time to get an initial estimate of the autocorrelation curve and its decay. If full decay is observed within 30 min, a time window of 1.5 h (3 times the decay constant) for XPCS scan can be considered. Exposure time is determined based on the SNR and sample dynamics. If dynamics is slower, higher exposure times can be considered. Although, for fast dynamics it is ideal to measure with shorter exposure time, but with a tradeoff with lower photon count of the speckle pattern.
5. **External Stimuli**: XPCS scans are usually collected for a range of external stimuli i.e. temperature, stress or fields. Thus, material characterization prior to XPCS beamtime is important in deciding the range of external stimuli values. An additional challenge for complex oxides is that near the phase transition regime, the magnetic, electronic or even structural ordering reflections usually decreases in intensity as the transformations occurs. The lower intensity can reduce the data quality and reliability of the measurements near the most interesting regime for material dynamics. Increased coherent flux for upcoming synchrotron sources will be key in capturing fluctuation dynamics near phase transitions.

3 Data Analysis and Fitting Methods

In order to analyze XPCS data and understand the dynamical fluctuations, intensity correlation functions are used. During a XPCS experiment, the size of X-ray beam is typically large compared to the size of particles (or domain size). Thus, the illuminated volume by X-rays contains a large number of particles (or domains). The evolution of macroscopic average of the particles can lead to very fast changes in the intensity of the scattered light. Autocorrelation functions are used to mathematically characterize these intensity fluctuations and obtain insights regarding motion of particles or fluctuations of domains or domain walls within the illuminated volume. For example, the diffusion rates of particles in a colloidal solution can be measured using DLS by calculating relaxation time from the fits of the autocorrelation function. In addition to correlation functions, waterfall plots or 'kymograph' are used to visually examine fluctuations in speckle pattern.

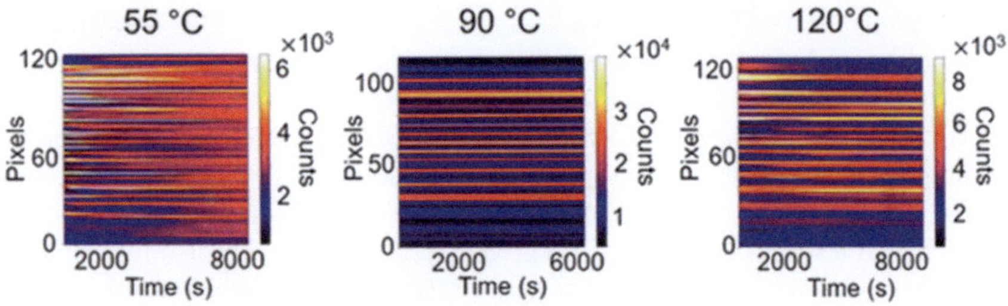

Fig. 11 Kymographs or waterfall plots for speckle pattern obtained from ferroelectric domains $BaTiO_3$ thin film. The waterfall plots are obtained by plotting the intensity as a function of time across a circular cut of the domain diffuse scattering near the Bragg Peak[25]

3.1 Kymographs or Waterfall Plots

Kymographs or waterfall plots are used to plot any linecut across the speckle pattern as a function of the measured time. In a speckle pattern, multiple speckles would be present, the evolution of which can be visually inspected for the entire time range across any line using waterfall plots. Figure 11 presents Kymographs for ferroelectric domains in $BaTiO_3$ thin film for different temperatures across the domain transformation, as discussed in Sect. 4 [25]. From the kymograph plots, it can be seen that the speckles for 90 °C remain stable throughout the three-hour measurement. Speckle fluctuations are observed for 55 and 120 °C, with speckles appearing and disappearing as a function of time. The fluctuations in the speckle pattern is the direct evidence of ferroelectric domain fluctuations near the domain transformation temperature (55 °C) and Curie temperature ($\sim$ 130 °C). Note that the unvarying speckle pattern also demonstrates the stability of the experimental setup. Recently, speckle tracking techniques have also been developed to quantify systematic shifts of the speckles to infer strain in highly disordered polymer [46].

3.2 Autocorrelation Function (g_2)

One of the first analysis step for XPCS is to calculate the auto-correlation function (g_2) to characterizes the dynamical fluctuations which is given by,

$$g_2(q, \Delta t) = \frac{\langle I(q,t)I(q,t+\Delta t)\rangle}{\langle I(q)\rangle^2} \tag{5}$$

where I is the intensity at a given q (or pixel) for a particular time. The average is performed over the time Δt, and thus g_2 is typically plotted as a function of time lag (and not 'real time' unless stated otherwise). For example, consider detector images taken at time Δt apart in Fig. 12 [47]. The auto-correlation function (g_2) for

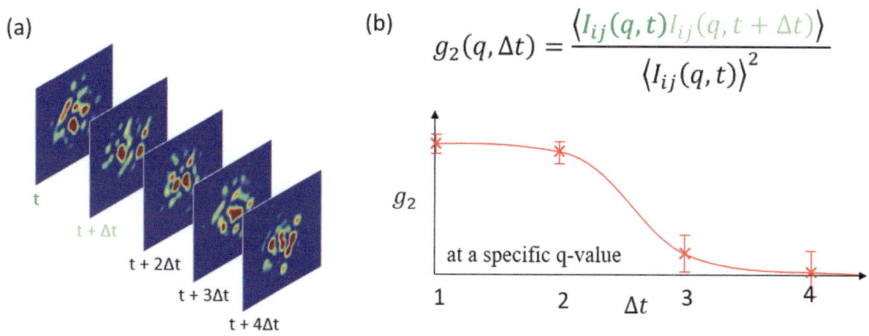

Fig. 12 Auto-correlation function (g_2): **a** series of 2D images are collected during XPCS scan, **b** graph showing the correlation as a function of time lag (Δt), known as a correlogram [47]

time lag $2\Delta t$ is calculated by averaging over all pairs which are $2\Delta t$ apart, i.e. t and $t + 2\Delta t$, $t + \Delta t$ and $t + 3\Delta t$, $t + 2\Delta t$ and $t + 4\Delta t$, to quantify how correlated the two speckle pattern are after time $2\Delta t$. In order to gain any timelag-resolved information, g_2 is found for many values of such time intervals. Also, since the speckle pattern is fully correlated with an exact copy of itself, the maximum possible correlation occurs for $\Delta t = 0$. If the value of autocorrelation function remains equal to this initial value throughout the measurement window, it implies that the sample is completely static. Otherwise, the function will decay from its initial value as the speckle pattern becomes uncorrelated with itself as the material changes from its initial state, providing fundamental insights into material dynamics. Thus, the XPCS setup needs to be extremely stable during the measurement window so that the decay in g_2 is due to sample dynamics and not drifts in the experimental setup.

Key things to note regarding auto-correlation functions are, (1) g_2 is plotted as a function of time lag, and thus provides insights into equilibrium fluctuations and processes independent of 'real time'. In order to quantify dependence on real time, two-time correlation as discussed below should be calculated. (2) Since all experimental auto-correlation functions must be found in discrete and finite time intervals, the number of points used to calculate each value decreases as Δt increases. Thus, auto-correlation for larger time delays is based on fewer and fewer data points, leading to nonphysical solutions in some cases due to averaging over too few values. For example, for a measurement window of 1.5 h, only one pair of speckle pattern is 1.5 h apart leading to one dataset with higher error bar. Such problems during the data analysis can manifest as an increase in g_2 above its initial value at high time lag values. For this reason, (3) auto-correlation functions are calculated only for $\Delta t <$ T/2, where T is the measurement window.

3.3 g_2 and Dynamical Structure Factor

Auto-correlation function, g_2, is related to the dynamical structure factor or the scattering function $S(q, \omega)$ [2, 3, 43]. $S(q, \omega)$ is given by,

$$S(q, \omega) = 1/2\pi \int_{-\inf}^{\inf} dt e^{-i\omega t} \langle \rho(q, 0)\rho(-q, t) \rangle \tag{6}$$

where $\rho(q, t)$ is the Fourier component of the electron density at time t, and ω represents frequency of the energy loss of the probe. The average can be either statistical or quantum mechanical over the material system. Measurement of $S(q, \omega)$ provides direct information about material properties and can also be measured using inelastic scattering. The frequency Fourier transform of dynamical structure factor ($S(q, \omega)$) is intermediate scattering function, $g_1(q, t)$ which is given by [2],

$$g_1(q, t) = 1/N \langle \rho(q, 0)\rho(-q, t) \rangle \tag{7}$$

here N is the total numbers of scatterers. On the other hand, the scattering intensity measured from the sample for a completely coherent X-ray beam is given by,

$$I(q, t) = C \langle \rho(q, 0)\rho(-q, t) \rangle \tag{8}$$

here C is constant related to experimental parameters such as the incident beam intensity etc. Note that the average here is over time t. This measured intensity on the detector is used to calculate the autocorrelation function, which is given by Eq. 5. It can be seen from Eqs. 5 and 8 that g_2 is second order in the intensity and fourth order in the electric fields. g_2 can also be written in terms of electric fields via Van Cittert–Zernike theorem, which is given by Seigert Relation (derivation can be found in Ref. [48]),

$$g_2(q, t) = 1 + g_1(q, t)/g_1(q, 0) = 1 + |F(q, t)|^2 \tag{9}$$

where $|F(q, t)|$ is normalized intermediate scattering function. Here t has been used instead of Δt in Eq. 5 for brevity. For a partially coherent beam, which is typical for most synchrotron experiments, this relationship is written as,

$$g_2(q, t) = 1 + \beta(q)|F(q, t)|^2 \tag{10}$$

where β is the q-dependent contrast factor that depends both on incident beam properties and experimental setup, and can vary from 0 (fully incoherent) to 1 (fully coherent). It is a measure of the signal to noise ratio, as it indicates how well the speckles can be resolved. For XPCS experiments, a higher value of β is desirable. Effects of partial coherence are discussed in detail in reference [8]. Note that in the above relationship, the phase of g_1 is undetermined. Different methods have been developed to obtain the phase information such as heterodyne XPCS experiments (where a reference beam is also used). For ergodic systems, g_2 can be calculated by averaging over a particular q-range (a set of pixels of an area detector) where the same dynamics can be assumed. Equation 10 is used to obtained intermediate scattering function from XPCS measurements, considering equilibrium flustrations.

3.4 g_2 and Signal to Noise Ratio

The signal to noise ratio in an XPCS study is given by [2, 49],

$$R_{sn} = \langle I \rangle \, \beta \sqrt{T \delta t N / g_2} \tag{11}$$

where $\langle I \rangle$ is the detector count rate, δt is the integration time of the detected signal, T is the duration of the experiment and N the number of independent detector channels averaged to obtain the signal. For a detailed discussion on how to improve signal to noise ratio in XPCS experiments, see Ref. [49].

3.5 Fitting Autocorrelation Function (g_2)

The next step in XPCS data analysis is to fit the intermediate scattering function obtained from g_2 (Eq. 10) to obtain meaningful information about the system. For most stochastic variables (those whose values are dictated by probabilities), the autocorrelation function decays as a single exponential function. Thus, in many cases, a characteristic time constant is obtained by fitting g_2 with a single exponential and obtaining τ which represents the time required for the autocorrelation function to decay from its initial value to 1/e times the initial value. For example, to investigate diffusion dynamics of collodial particles in a solution [50, 51], the ISF was fitted using a single exponential to obtain relaxation timescales of collodial particles undergoing Brownian motion,

$$F(q, t) = e^{-Dq^2 t} \tag{12}$$

For hydrodynamic diffusion, the relaxation time constant is given by $\frac{1}{Dq^2}$, which can be used to determine the diffusion constant (D) of particles in a solution. This in turn can be used to estimate the hydrodynamic radius of the particle using Stokes–Einstein relation.

However, for many materials including soft materials, and recently quantum materials, the decay for ISF is found to occur over either a much longer or shorter time interval, deviating from the single exponential form. In this case, ISF is modeled by considering a Kohlrausch–Williams–Watts (KWW) modified exponential form which is given by,

$$F(q,t) = e^{-(t/\tau)^{\gamma}} \tag{13}$$

here τ is the decay constant and γ (sometimes also referred as β and not to be confused with speckle contrast which is also sometimes referred as β in the literature) is the exponent. These two parameters can be dependent or independent of q based on the material system. For values of $0 < \gamma < 1$, considered as a stretched exponent, the system is considered to be glassy, with little spatial ordering and long-timescale dynamics. On the other hand, for values of $\gamma > 1$, considered as a compressed exponent, the system is considered as jammed, or in a state of structural arrest. Recently, $\gamma > 1$ has been observed for a variety of complex oxides including antiferromagnetic fluctuations in Cr [17], orbital dynamics in magnetite [9, 21] and ferroelectric domain fluctuations as shown in Fig. 16 [25].

Note that a compressed exponential represents a faster decay, which seems to be counter intuitive to the idea of a jammed system in which the system is largely static. However, it can be understood in the context that a jammed system is one in which there are long periods of time with little to no motion, and short periods of time with lots of motion. For example, consider being packed onto a crowded bus. In order to get off the bus at a given stop, instead of pushing your way through the crowd, it is easiest to wait for those closest to the door to file off, then follow the string of people who leave the bus. Fit parameters τ and γ are plotted as a function of temperature (or other external stimuli) and sometimes within portions of the same scan (see discussion below on non-equilibrium dynamics) to investigate material dynamics. In addition to τ and γ, speckle contrast $\beta(q)$ intensity, full width half maximum (FWHM) and Bragg peak position are plotted as function of temperature or portions of the same scan to check for any experimental variability and investigate material behavior. Note that different models in addition to KWW model have been used and details on different models for fitting ISF can be found in References [2, 39, 42].

3.6 q-Dependence

In order to obtain additional information regarding fluctuations dynamics, relaxation time (τ_0) and the KWW exponent can be plotted as a function of reciprocal wavevector (q). This can be achieved by measuring speckle pattern for different q-vectors or by sectioning the Bragg peak in concentric regions of interest from the center. Figure 13 plots the relationship between timescales versus the lengthscales along with different processes occurring during phase transition. The green and red lines in the figure indicate diffusive processes with a slope of ~ 2. Processes with slope lower than 2 are faster than diffusion as indicated by blue and grey lines, which represent speed of sound and light respectively, while processes with slope higher than 2 will be slower than diffusion (not shown) [42].

While q-dependence has not yet been reported for complex oxides, strong q-dependence has been observed in polymers, gels and glasses. Figure 14a plots the

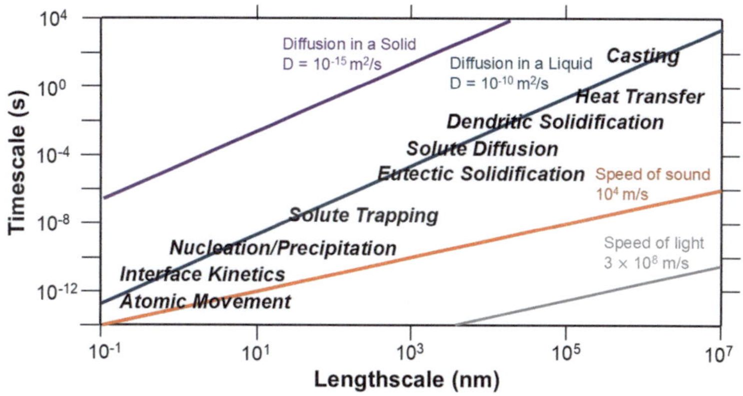

Fig. 13 Relationship between timescales and lengthscales for different processes during phase transition. Adapted from Reference [42]

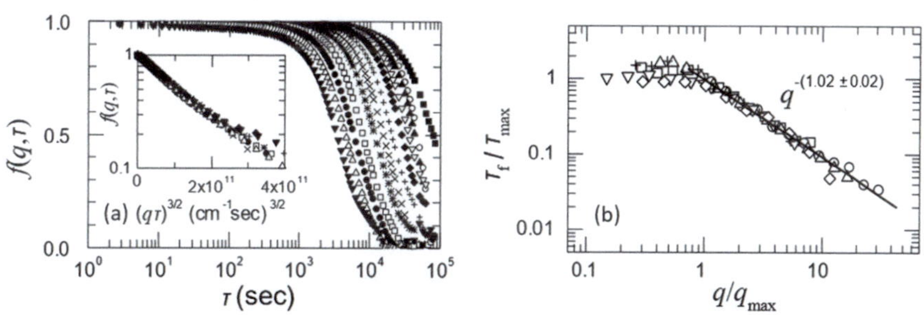

Fig. 14 Q dependence of the decay constant (τ) of a collodial gel system: **a** $F(q, \tau)$ for different q-values. From right to left, q varies from 249 to 6745 cm^{-1}. Inset plots $F(q, \tau)$ versus $(q\tau)^{3/2}$ for $q > q_{max}$, **b** the scaled decay time of the dynamic structure factor as a function of the scaled scattering vector [52]

$F(q, \tau)$ for colloidal gels for different q-values ranging from 249 to 6745 cm^{-1} measured using DLS [52]. All the measured data was collapsed on to a single curve that plots the scaled decay constant as a function of the scaled q-vector as shown in Fig. 14b. It was observed that decay time constant is inversely dependent on q ($\tau \propto q^{-1}$) for $q > q_{max}$, which rules out diffusive motion and instead suggests a drift mechanism. Based on the measured q and τ dependence of $F(q, \tau)$, Cipelletti et al. [52] proposed a model based on syneresis of the gel whereby the gel shrinks locally, and the resulting inhomogeneity results in random regions of higher deformation which in turn act as a force dipole on nearby regions.

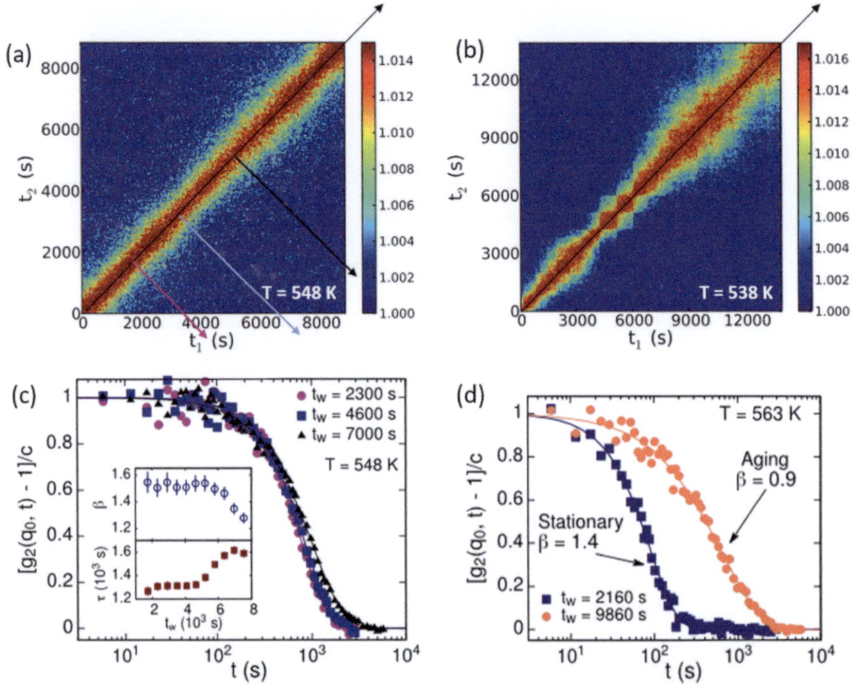

Fig. 15 Two-time correlation functions (TTCF) for a bluk metallic glass: **a** TTCF at 548 K showing equilibrium behavior as the width of TTCF stays the same throughout the scan, and **b** TTCF at 538 K showing non-equilibrium dynamics as the width of the TTCF changes for different aging times. Long black arrow indicates $t_1 = t_2$ line which is fully correlated. Perpendicular cuts (pink, blue and short black arrow) to this long black arrow shows the correlation function with increasing 'real time' or 'aging time' or 'wait time'. **c** Correlation function calculated for different wait times including 2300, 4600, 7000 s which are represented in the TTCF by pink, blue and black arrow respectively. Inset shows fitted decay constant τ and exponent β. **d** Correlation functions calculated at 563 K for wait times of 2160 and 9860 s, clearly show distinct dynamics at two different wait times [53]

3.7 Non-equilibrium Dynamics

The discussion in this section has so far focused on equilibrium dynamics, where the correlation function g_2 (Eq. 5) is plotted as time lag and not 'real time'. However, in non-equilibrium systems, such as quenched metallic glasses, soft matter systems, or complex oxides such as magnetite, this assumption is not valid. and the relaxation time constant is a function of 'real time' or 'aging time' or 'wait time'. The non-equilibrium dynamics is quantified using two-time correlation function (TTCF), $g_2(q, t_1, t_2)$, which is given by,

$$g_2(q, t_1, t_2) = \frac{\langle I(q, t_1) I(q, t_2) \rangle}{\langle I(q, t_1) \rangle \langle I(q, t_2) \rangle} \tag{14}$$

where $I(q, t_1)$ and $I(q, t_2)$ represent intensity of speckle patterns measured at two different times, t_1 and t_2. TTCF reflects the statistical similarity between average sample behavior measured at times t_1 and t_2 and is the time-resolved version of the autocorrelation function, g_2 (Eq. 5). TTCF and one-time correlation function for $Pd_{43}Cu_{27}Ni_{10}P_{20}$ bulk metallic glass are shown in Fig. 15 for different temperatures in the vicinity of the glass transition temperature (T_g = 563 K) [53]. Note that both axis in Fig. 15a, b are real time t_1 and t_2. The $t_1 = t_2$ line (long black arrow, the central diagonal) is fully correlated as that is the same speckle pattern correlated with itself. This center diagonal, where the two-time correlation function has the highest intensity, corresponds to the elapsed time of the measurements i.e. wait time. The lines orthogonal to the central diagonal shows how the correlation function decays with the increasing wait time. The correlation function decreases as the difference between t_1 and t_2 increases i.e. off-diagonal. The width of this center diagonal profile is proportional to τ, the characteristic time scale (decay constant) of the correlation function for equilibrium dynamics. For non-equilibrium dynamics τ values varies with wait time as discussed below.

For example, in Fig. 15a pink, blue and black arrows can be used to extract the one-time correlation function at different wait times [53]. Figure 15c shows the one-time correlation function obtained from TTCF for different wait time (t_w) of 2300, 4600 and 7000 s. One-time correlation functions were fitted using the KWW model to obtain τ and β (same as γ for Eq. 13) and are shown in the inset. No variation in either the width of center diagonal in TTCF or the decay constant for one-time correlation function for different wait times is observed. Thus, the dynamics is considered to be homogeneous throughout the measurement i.e. equilibrium dynamics. For equilibrium dynamics, the autocorrelation function (g_2, Eq. 5) can be used to quantify the characteristic decay time and the exponent. In fact, g_2 is an average of all the lines orthogonal to the central diagonal in TTCF i.e. g_2 is an average of all one-time correlation cuts in TTCF. Overall, g_2 is a function of time lag Δt, while two-time correlation function is a function of wait time.

On the other hand, Fig. 15b is an example of non-equilibrium dynamics where different sections show slow or fast decay along the line orthogonal to the center diagonal. Figure 15d plots the one-time correlation function across such sections showing the drastic difference in one-time correlation function as a function of wait time (t_w). Note that the temperatures for Fig. 15b, d are different, indicating that non-equilibrium dynamics are observed for both of these temperatures near glass transition. As the system ages (with increasing t_w), the decay constant is found to be increasing on average, as indicated by the width of the TTCF. However, there are points where the two-time autocorrelation function pinches off. This shape is indicative of a sudden and discontinuous rearrangement followed by a quick return to the original relaxation timescale. Such rearrangements are a signature of local avalanches in a jammed system [54]. Note that careful analysis should be done to rule out the fact that these sudden changes are not due to variable X-ray beam intensity or any drifts in the experimental setup during the measurements. Based on the measured two-time correlation function, Evenson et al. [53] proposed a mechanism based on aging as the consequence of gradual release of atomic-level stresses. These experimental observations are also in agreement with the model proposed by Bouchard and Pitard [55], where release of internal stress occurs due to structural rearrangements. Elastic relaxation is driven by localized "micro-collapses" of group of particles which result in the occurrence of force dipoles which can trigger subsequent collapses in the neighboring regions. Micro-collapse or avalanche based dynamics has been found in different materials, for example, martensitic transformation of Cobalt [54], crystallization of a hardsphere glass [56], and colloidal gels [57].

4 Recent XPCS Studies on Complex Oxides

In this section, we discuss two recent XPCS studies on complex oxides focusing on the ferroelectric domain fluctuations in low-strain $BaTiO_3$ thin films near domain transformation temperature and Curie temperature, and charge and orbital order fluctuations in magnetite near the Verwey transition.

4.1 Ferroelectric Domain Fluctuations in $BaTiO_3$

XPCS studies of domain fluctuations in ferroelectric thin films have been conducted by accessing to domain diffuse scattering near the Bragg Peak [24, 25]. Li et al. [25] investigated a/c domain fluctuations in a low-strain (001) $BaTiO_3$ (80 nm) thin film epitaxially grown on $SrRuO_3$ (6 nm)/$NdScO_3$ substrate for a temperature range covering domain transformation temperature (T_R) and Curie temperature (T_C $\sim$ 130 °C). Low-strain $BaTiO_3$ shows a/b domain pattern at room temperature, and undergoes a transformation to a/c domain pattern at T_R = 55 °C [58, 59]. Figure 16a presents the scattering geometry used to access the a/c domain diffuse scattering near the (001) Bragg reflection non-resonantly. The domain diffuse scattering is due to the periodic arrangement of a/c domains resulting in ordered domain walls [60, 61]. Figure 11 already presented the kymographs or "waterfall" plots, depicting the evolution of intensity along a circular cut for T = 55, 90, and 120 °C. The variation of speckle intensity in the waterfall plots for different temperatures directly demonstrates the strong temperature dependence of the domain flutuations.

Figure 16b presents the ISF calculated from the autocorrelation function, g_2 (Eqs. 5 and 10), for selected temperatures during the heating cycle [25]. No significant q-dependence was observed for these measurements. Additionally, non-equilibrium dynamics was observed from the TTCF plots (see supplementary information of [25]). Figure 16c presents the fit parameters, τ and β, obtained by fitting the ISF using KWW model (Eq. 13) as a function of temperature. The inset of Fig. 16c shows the exponent, β, to be greater than 1 for all the measured temperatures, indicative of jamming behavior. Note that β greater than 1 has been reported for ferroelectric $PbTiO_3$/$SrTiO_3$ superlattice thin films [24], materials exhibiting charge density waves [17], and charge and orbital ordering in magnetite [9]. Figure 16d shows the

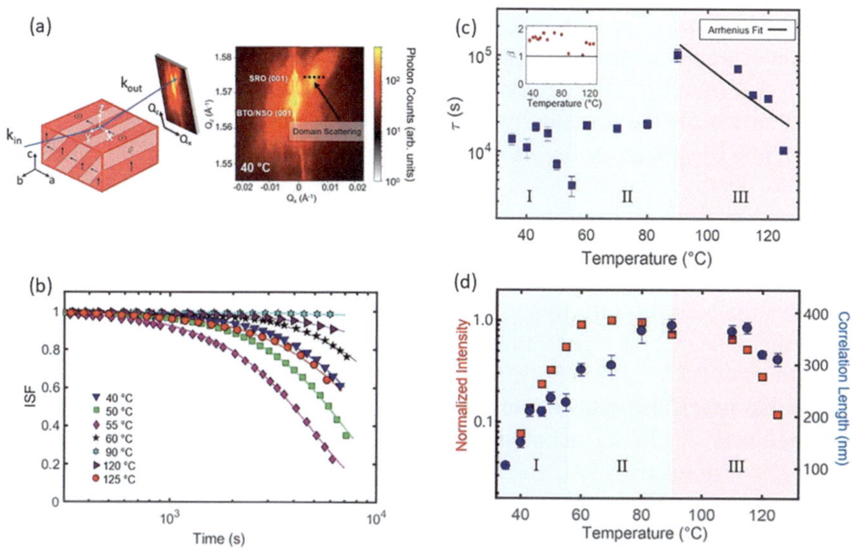

Fig. 16 Ferroelectric domain fluctuations in $BaTiO_3$ thin film: **a** X-ray scattering geometry and the 2D detector image showing a/c diffuse domain scattering near the (001) Bragg reflection. **b** ISF and fits to the ISF for selected temperatures. **c** Characteristic domain fluctuation timescales τ as a function of temperature obtained from fits to the ISF. Three different regimes as discussed in the text are highlighted. Inset shows the exponent β as a function of temperature obtained from fits to the ISF. **d** Normalized intensity and correlation length obtained from the a/c domain diffuse scattering [25]

normalized intensity (red square symbols), and correlation length, λ (blue circle symbols), of the a/c domain diffuse scattering.

Three distinct temperature regimes (Fig. 16c, d) were identified based on the evolution of the decay constant, intensity and correlation length [25]. In regime I, as T_R is approached from room temperature, integrated intensity and correlation length increase monotonically indicating a transformation from the a/b to the a/c domain pattern. Although, the decay constant, τ, during the initial phase of regime I (from 35 to 45 °C) shows no obvious temperature dependence. Lack of temperature dependence of τ was attributed to the growth of a/c domains in a predominantly a/b matrix, which results in similar boundary conditions in this temperature range. Only later in regime I, with a further increase in temperature (from 45 to 55 °C), τ decreases as the transformation to the a/c domains continues, resulting in a system with predominantly a/c domains.

In regime II, a slow-down of domain fluctuations was observed with a strong increase in τ from 55 to 60 °C and 80 to 90 °C. The first jump from 55 to 60 °C was related to the completion of a/b to a/c domain transformation also marked by no further change in intensity in regime II. Although, the correlation length continued to increase and only stabilized in a temperature range of 80–90 °C, which is concomitant with the second jump in τ. This increase in correlation length was attributed to improved ordering of periodic a/c domains likely achieved by overcoming pinning due to defects or other pinning sites including the presence of a small fraction of remnant a/b domains [59]. Thus, the second jump could be due to complete stabilization of a/c domains and the final disappearance of a/b domains.

In regime III, τ decreases as the Curie temperature is approached ($T_C \sim 130\,°C$) and more thermal energy is injected. The intensity and correlation length also decrease, signaling a reduction in a/c domains and growth of the paraelectric phase 20 °C below the Curie temperature. Standard thermally activated Arrhenius model fitting resulted in an activation energy of 0.67 ± 0.27 eV. This value is similar to the activation barrier for the migration of oxygen vacancies (0.8 eV) [62], as well as within the same range for generation and annihilation of topological defects (activation barrier of 0.35 ± 0.21) [24]. Thus, both of these processes were proposed for fluctuations near Curie temperature. Note that a slow down of τ was seen for datasets measured near $\sim$ 125 °C (unpublished), although the SNR decreases strongly due to a decrease in intensity as T_C is approached. Development of fully coherent synchrotron sources will enable higher intensity and sufficient SNR to reliably capture dynamics near critical temperatures.

4.2 Charge-Orbital Order Fluctuations in Magnetite Below the Verwey Transition

Charge-orbital ordering fluctuations in magnetite have been measured near Verwey transition, by combining XPCS and resonant elastic scattering at Fe L-edge and O K-edges, to investigate whether fluctuations are due to structural domain reorientation or purely electronic in nature [9, 21]. The experimental schematic to access (001/2) superlattice peak resonantly either at Fe L-edge or O K-edge is shown in Fig. 17. Fe L_3 edge is sensitive to Fe 3d orbital via excitation from the core to unoccupied valence states ($2p_{3/2}$ to 3d transitions) and directly probes the B-site Fe^{2+} t_{2g} orbital structure [63]. Oxygen K-edge (or in this case slightly before the absorption edge to maximize resonant scattering) probes the O 2p orbital via excitation from core to unoccupied valence states (1s to 2p transitions) [64]. Note that both the oxygen K-edge and iron L-edge resonant signals are only sensitive to the insulating phase. As seen in Fig. 17b, d a drop in total intensity occurs as metallic domains and grow in the vicinity of metal-insulation transistion.

Figure 17c shows the decay constant (τ) as a function of temperature obtained from the KWW fits to the ISF. A decrease in fluctuation time constant is observed as temperature is increased from 20 K (Arrhenius dependence), but a slow down is detected near the vicinity of the phase transition. The slowdown in decay constant was initially related to the phase separation of metallic and insulating domains as indicated by the lowering of correlation length in Fig. 17b [9].

Measurements at the oxygen K-edge later revealed no fluctuations as shown in two-time correlation plots in Fig. 18 [21]. Surprisingly, the two-time correlation plots

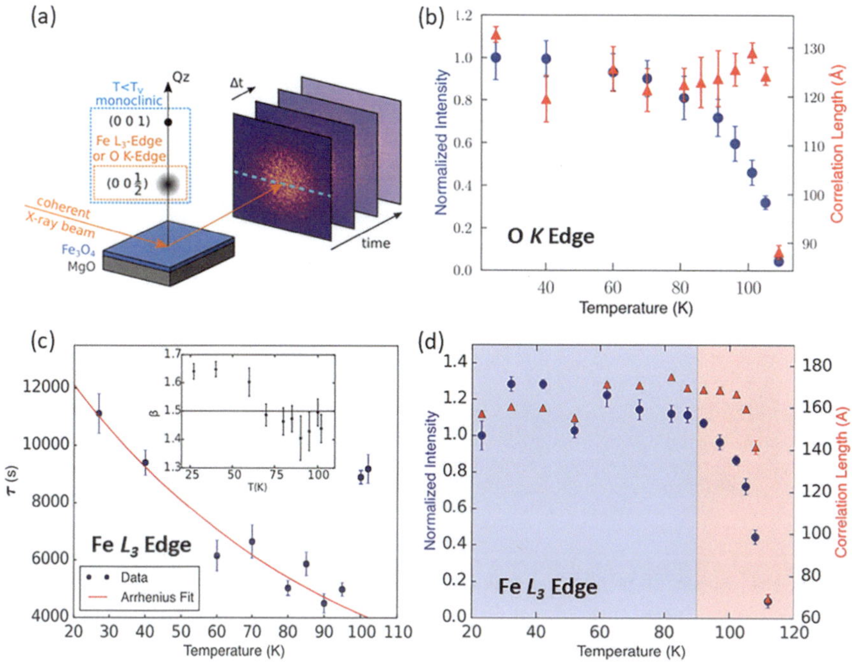

Fig. 17 Charge-orbital order fluctuations in magnetite below the Verwey transition: **a** X-ray scattering geometry used to access (001/2) superlattice peak resonantly either at Fe L_3 edge or oxygen K edge. **b** Normalized intensity and correlation length calculated from resonant scattering at the Fe L-edge. **c** Decay constant τ and β (inset) as a function of temperature obtained from fits to the ISF. **d** Normalized intensity and correlation length calculated from resonant scattering at the oxygen K edge [9, 21]

at the Fe L_3 edge show non-equilibrium behavior while a static signal is observed for the oxygen K-edge. The horizontal and vertical dashed lines mark the stability of the sample and experimental setup. Oxygen K-edge measurements clearly show that the correlation value for most of the plot is approximately the same as the $t_1 = t_2$ line (as discussed above this central diagonal line represents full correlation), indicating a static behavior. No fluctuations at the oxygen K-edge rule out structural fluctuations or phase separation of metallic and insulating domains as the origin of fluctuations at the Fe the L-edge. If there was enough thermal energy present in the system to drive structural fluctuations between metallic and insulating phases at each temperature, fluctuations should have been observed at both the oxygen K-edge and Fe L-edge, which is clearly not the case.

Hua et al. [21] thus proposed the origin of the dynamic speckle pattern at Fe L-edge to be electronic in nature. They proposed an extremely stable oxygen 2p orbital network in the low-temperature monoclinic phase where there is an absence of ligand-to-metal charge transfer dynamics in equilibrium. They postulated that the thermally active fluctuations observed for the Fe t_{2g} orbital domains are due to

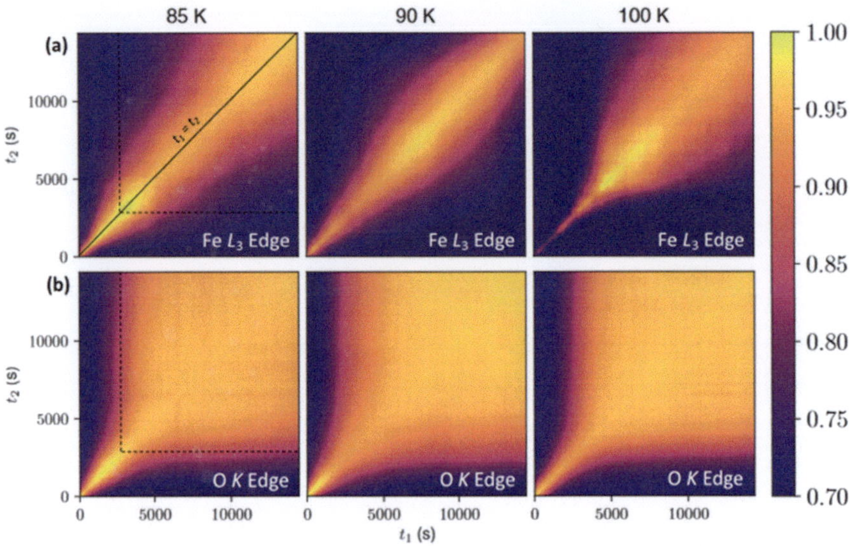

Fig. 18 Normalized TTCF for magnetite below Verwey transition: TTCF were measured at the **a** Fe L_3 edge and **b** oxygen K edge at 85, 90 and 100 K. The approximate beginning of stable equilibrium is marked by dashed black lines [21]

the delocalization of the minority spin electron between adjacent B-site iron cations within the trimeron network, which is in alignment with Ref. [65]. Thus, dynamical fluctuation studies using XPCS were able to point out key differences in Fe and Oxygen charge and orbital ordering which have been considered strongly hybridized based on the static measurements.

In this chapter, a summary of XPCS techniques including experimental details, data analysis and XPCS studies on complex oxide was presented. In the upcoming future, development of coherent synchrotron sources as well as new advances in all aspects of experimental setup, beamline setups, detectors and analysis tools will enable XPCS measurements of wide variety of material systems. These developments will allow us to capture the regions near critical points of transition with higher spatial resolution. Development of XPCS techniques at XFEL's will provide access to femtosecond to nanosecond timescales. Overall, these new advances will enable fundamental insights into material dynamics at the ultrasmall and ultrafast frontier.

Questions and Problems (Coherence X-Ray Techniques for Nanoscale Characterization of Complex Oxides)

Q 4.1. **Coherence and coherence lengths**

i. Experiments exploiting the transverse coherence of X-ray sources are typically conducted using instruments that are located a long distance from a spatially small X-ray source. Estimate the transverse coherence of the X-ray beam typically used in laboratory X-ray experiments. A typical Cu Kα laboratory source could have a source-to-sample distance of $R = 1$ m and a size $w = 0.5$ mm. What is the transverse coherence length of this source? Explain why it is possible on such a source to measure the thickness of a 100 nm thin-film layer.

ii. The longitudinal coherence length becomes important when beams with different total distances of propagation undergo interference. This case arises in asymmetric diffraction geometries. What longitudinal coherence length is required to observe the intensity oscillations of a thin film with a thickness of 100 nm using a particular X-ray reflection with an incident angle of 10° and exit angle of 40° in the geometry shown below? *Hint: Consider the difference in the distance traveled by X-rays reflected from near the top and both surfaces.*

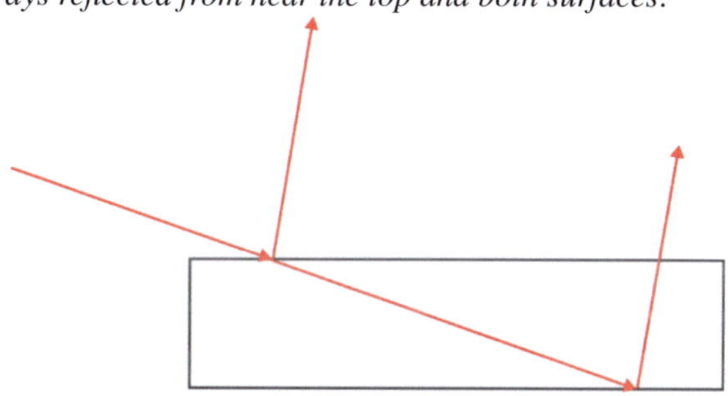

iii. The X-ray beamline shown in Fig. 9 offers $\Delta\lambda/\lambda = 3 \times 10^{-4}$. Calculate the maximum thickness of the film from which thickness oscillations can be observed at this beamline using the diffraction geometry shown in part (ii).

Q 4.2. **Coherence experiments with soft X-rays often impose a different set of constraints than with hard X-rays**

i. Compute the transverse coherence length for X-rays with photon energy 800 eV and 20 keV at a light source with a source-to-sample distance of 80 m and source size of 20 μm. Assume that the X-ray beam has an overall diameter of 500 μm at the sample. Is either of these X-ray beams fully coherent?

ii. How close to a 20 μm-size X-ray source could you move the sample while still having a transverse coherence length of 100 μm? Does this explain why soft X-ray beamlines are often shorter than hard X-ray coherence beamlines?

Q 4.3. **Coherent volume versus diffraction volume**

i. Consider a bulk Fe sample. Calculate the diffraction volume (transverse and longitudinal) for photon energies of 800 eV and 20 keV with a source size of 20 μm considering normal incidence. Penetration depth at these energies can be found on CXRO website (www.cxro.lbl.gov).

ii. The hard and soft X-ray beamlines shown in Figs. 9 and 10 offer $\Delta\lambda/\lambda = 3 \times 10^{-4}$ and 2×10^{-3}, respectively. Calculate the longitudinal coherence for photon energy of 20 keV and 800 eV for these two beamlines.
iii. Using the longitudinal coherence calculated in part (ii) and transverse coherence length the Question 2(ii), comment on the relative size of diffraction volume versus coherent volume for hard and soft X-ray energy range, and how that influences the speckle contrast in X-ray photon correlation spectroscopy (XPCS) experiments.
iv. XPCS measurements for complex oxides are typically performed in the Bragg geometry and not in normal incidence. Consider X-ray beam incident at 15° angle to satisfy the Bragg condition. Calculate the diffraction volume (transverse and longitudinal) for this geometry for 800 eV and 20 keV with a source size of 20 μm. Comment on how the diffraction volume calculated for Bragg geometry compares to the coherence volume for the hard and soft X-ray energy range, and its influence on the speckle contrast.

Q 4.4. XPCS experiments require that the speckle pattern be acquired with sufficient angular resolution to distinguish each speckle. Typically experiments would require that at least one pixel (of the detector) samples one speckle. How far does a detector with pixel size of 50 μm need to be from the sample to resolve the speckle pattern generated by an X-ray beam with a diameter of $\sigma = 10$ μm and wavelength 0.1 nm? What distance would be required with a detector with a pixel size of 200 μm?

Q 4.5. Coherent diffraction imaging relies on an interpretation of the intensity distribution in reciprocal space to form a real-space image of an object. One limit to the spatial resolution of this approach is set by the total size of the detector. What spatial resolution is in principle possible with a 10 cm-wide detector located 1 m from a sample in a coherent diffraction imaging (CDI) experiment conducted with 0.1 nm wavelength X-ray radiation?

Q 4.6. XPCS experiments have a signal-to-noise ratio (SNR) that depends on the intensity of the incident X-ray beam. The XPCS analysis has an SNR given by Eq. 11. Derive an equation for the time resolution τ as a function of intensity for a set value of SNR. By what factor does τ change if the intensity is increased by a factor of 10?

Q 4.7. Analysis of XPCS datasets is usually performed using autocorrelation functions i.e. g_2 or two-time correlation functions (TTCF). Consider the plot below which shows the scattered intensity versus time plot obtained during an XPCS scan for a particular temperature [43].

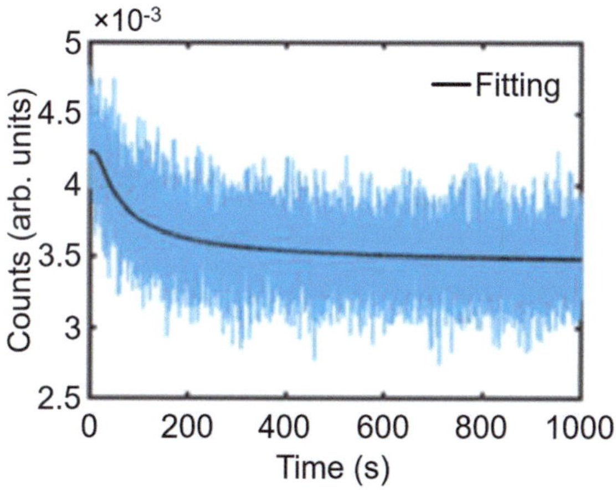

i. List one sample and one experimental factor which could lead to the variation of the scattered intensity from the sample during an XPCS scan.
ii. Usually for intensity variations, kymographs as well as TTCF or one-time correlation functions at different wait times are verified. Identify three wait times and time windows on this plot, for which one-time correlation function can be calculated to investigate if the intensity variation and changes in the correlation function could be related.
iii. Consider a scenario where no variation in one-time correlation function is observed, and autocorrelation function, g_2, is found to be 1 throughout the scan. How would that influence your choice of measurement window and exposure time?
iv. On the other hand, if the autocorrelation function, g_2, is stable at the maximum contrast observed at the beamline, what would that imply? Would that influence your choice of measurement window and exposure time?

References

1. M. Sutton, X-ray intensity fluctuation spectroscopy, chap. 9, in *Neutron and X-Ray Spectroscopy* (Springer, 2006), pp. 297–318
2. S.K. Sinha, Z. Jiang, L.B. Lurio, X-ray photon correlation spectroscopy studies of surfaces and thin films. Adv. Mater. **26**(46), 7764–7785 (2014)
3. G. Grübel, A. Madsen, A. Robert, X-ray photon correlation spectroscopy (XPCS), in *Soft Matter Characterization*, ed. by R. Borsali, R. Pecora (Springer, Dordrecht, 2008), pp. 953–995. ISBN: 978-1-4020-4465-6
4. A.L. Schawlow, Laser light. Sci. Am. **219**, 120 (1968)
5. https://zeiss-campus.magnet.fsu.edu/tutorials/coherence/indexflash.html
6. F. Van Der Veen, F. Pfeiffer, Coherent X-ray scattering. J. Phys. Condens. Matter **16**(28), 5003–5030 (2004)
7. A.R. Sandy, L.B. Lurio, S.G.J. Mochrie, A. Malik, G.B. Stephenson, J.F. Pelletier, M. Sutton, Design and characterization of an undulator beamline optimized for small-angle coherent X-ray scattering at the advanced photon source. J. Synchrotron Radiat. **6**(6), 1174–1184 (1999)

8. S. Sinha, M. Tolan, A. Gibaud, Effects of partial coherence on the scattering of X rays by matter. Phys. Rev. B Condens. Matter Mater. Phys. **57**(5), 2740–2758 (1998)
9. R. Kukreja, N. Hua, J. Ruby, A. Barbour, W. Hu, C. Mazzoli, S. Wilkins, E.F. Fullerton, O.G. Shpyrko, Orbital domain dynamics in magnetite below the Verwey transition. Phys. Rev. Lett. **121**, 177601 (2018)
10. N. Hua, Nanoscale studies of strongly correlated electron materials with coherent X-rays. Ph.D. thesis, University of California San Diego, 2020
11. J. Miao, D. Sayre, H.N. Chapman, Phase retrieval from the magnitude of the Fourier transforms of nonperiodic objects. J. Opt. Soc. Am. A **15**(6), 1662 (1998)
12. J. Miao, D. Sayre, On possible extensions of X-ray crystallography through diffraction-pattern oversampling. Acta Crystallogr. Sect. A Found. Crystallogr. **56**, 596–605 (2000)
13. B. Ruta, G. Baldi, Y. Chushkin, B. Rufflé, L. Cristofolini, A. Fontana, M. Zanatta, F. Nazzani, Revealing the fast atomic motion of network glasses. Nat. Commun. **5** (2014). ISSN: 2041-1723
14. A. Fluerasu, A. Moussaïd, A. Madsen, A. Schofield, Slow dynamics and aging in colloidal gels studied by X-ray photon correlation spectroscopy. Phys. Rev. E Stat. Nonlinear Soft Matter Phys. **76**(1), 3–6 (2007). ISBN: 1539-3755
15. A. Das, P.M. Derlet, C. Liu, E.M. Dufresne, R. Maaβ, Stress breaks universal aging behavior in a metallic glass. Nat. Commun. **10**(1), 1–9 (2019)
16. S.W. Chen, H. Guo, K.A. Seu, K. Dumesnil, S. Roy, S.K. Sinha, Jamming behavior of domains in a spiral antiferromagnetic system. Phys. Rev. Lett. **110**(21), 1–5 (2013)
17. O.G. Shpyrko, E.D. Isaacs, J.M. Logan, Y. Feng, G. Aeppli, R. Jaramillo, H.C. Kim, T.F. Rosenbaum, P. Zschack, M. Sprung, S. Narayanan, A.R. Sandy, Direct measurement of antiferromagnetic domain fluctuations. Nature **447**, 68–71 (2007)
18. J.G. Ulbrandt, M.G. Rainville, C. Wagenbach, S. Narayanan, A.R. Sandy, H. Zhou, K.F. Ludwig Jr., R.L. Headrick, Direct measurement of the propagation velocity of defects using coherent X-rays. Nat. Phys. **12**(8), 794–799 (2016)
19. M.G. Rainville, C. Wagenbach, J.G. Ulbrandt, S. Narayanan, A.R. Sandy, H. Zhou, R.L. Headrick, K.F. Ludwig, Co-GISAXS technique for investigating surface growth dynamics. Phys. Rev. B Condens. Matter Mater. Phys. **92**(21), 1–8 (2015)
20. J.J. Turner, K.J. Thomas, J.P. Hill, M.A. Pfelfer, K. Chesnel, Y. Tomioka, Y. Tokura, S.D. Kevan, Orbital domain dynamics in a doped manganite. New J. Phys. **10** (2008). arXiv: 0709.3558. ISSN: 1367-2630
21. N. Hua, J. Li, S.B. Hrkac, A. Barbour, H. Wen, C. Mazzoli, S. Wilkins, R. Kukreja, E.E. Fullerton, O.G. Shpyrko, Discerning element and site-specific fluctuations of the charge-orbital order in Fe_3O_4 below the Verwey transition. Phys. Rev. Mater. **7**(1), 014413 (2023)
22. Y. Shen, G. Fabbris, H. Miao, Y. Cao, D. Meyers, D.G. Mazzone, T. Assefa, X.M. Chen, K. Kisslinger, D. Prabhakaran, A.T. Boothroyd, J.M. Tranquada, W. Hu, A.M. Barbour, S.B. Wilkins, C. Mazzoli, I.K. Robinson, M.P.M. Dean, Charge condensation and lattice coupling drives stripe formation in nickelates. Phys. Rev. Lett. **126**(17), 1–7 (2021). arXiv: 2104.00082
23. X.M. Chen, V. Thampy, C. Mazzoli, A.M. Barbour, H. Miao, G.D. Gu, Y. Cao, J.M. Tranquada, M.P.M. Dean, S.B. Wilkins, Remarkable stability of charge density wave order in $La_{1.875}Ba_{0.125}CuO_4$. Phys. Rev. Lett. **117**(16), 167001 (2016)
24. Q. Zhang, E.M. Dufresne, P. Chen, J. Park, M.P. Cosgriff, M. Yusuf, Y. Dong, D.D. Fong, H. Zhou, Z. Cai, R.J. Harder, S.J. Callori, M. Dawber, P.G. Evans, A.R. Sandy, Thermal fluctuations of ferroelectric nanodomains in a ferroelectric-dielectric $PbTiO_3/SrTiO_3$ super lattice. Phys. Rev. Lett. **118**(9), 097601 (2017)
25. J. Li, L. Zhong, R. Jangid, F.N.U. Meera, G. Rippy, K.B. Ainslie, C. Kohne, A.S. Everhardt, B. Noheda, Y. Zhang, A. Fluerasu, S. Matzen, R. Kukreja, Domain fluctuations in a ferroelectric low-strain $BaTiO_3$ thin film. Phys. Rev. Mater. **4**, 114409 (2020)
26. M.H. Seaberg, B. Holladay, S.A. Montoya, X.Y. Zheng, J.C.T. Lee, A.H. Reid, J.D. Koralek, L. Shen, V. Esposito, G. Coslovich, P. Walter, S. Zohar, V. Thampy, M.F. Lin, P. Hart, K. Nakahara, R. Streubel, S.D. Kevan, P. Fischer, W. Colocho, A. Lutman, F.J. Decker, E.E. Fullerton, M. Dunne, S. Roy, S.K. Sinha, J.J. Turner, Spontaneous fluctuations in a magnetic Fe/Gd skyrmion lattice. Phys. Rev. Res. **3**(3), 1–8 (2021)

27. M.H. Seaberg, B. Holladay, J.C.T. Lee, M. Sikorski, A.H. Reid, S.A. Montoya, J.D. Koralek, G. Coslovich, S. Moeller, W.F. Schlotter, R. Streubel, S.D. Kevan, E.E. Fullerton, J.L. Turner, F. Decker, S.K. Sinha, S. Roy, J.J. Turner, Nanosecond X-ray photon correlation spectroscopy on magnetic skyrmions. Phys. Rev. Lett. **119**, 067403 (2017)
28. W. Roseker, S.O. Hruszkewycz, F. Lehmkühler, M. Walther, H. Schulte-Schrepping, S. Lee, T. Osaka, L. Strüder, R. Hartmann, M. Sikorski, S. Song, A. Robert, P.H. Fuoss, M. Sutton, G.B. Stephenson, G. Grübel, Towards ultrafast dynamics with split-pulse X-ray photon correlation spectroscopy at free electron laser sources. Nat. Commun. **9**(1), 1704 (2018)
29. F. Dallari, M. Reiser, I. Lokteva, A. Jain, J. Möller, M. Scholz, A. Madsen, G. Grübel, F. Perakis, F. Lehmkühler, Analysis strategies for MHz XPCS at the European XFEL. Appl. Sci. **11**(17) (2021)
30. T. Fujita, Y. Sun, H. Li, T.J. Albert, S. Song, T. Sato, J. Moesgaard, A. Cornet, P. Sun, Y. Chen, M. Mo, N. Amini, F. Yang, A. Makareviciute, G. Coleman, P. Lucas, J.P. Embs, M. Sprung, V. Esposito, J. Vila-Comamala, N. Wang, T. Mamyrbayev, C. David, P. Zalden, J. Hastings, B. Ruta, P. Fuoss, K. Sokolowski-Tinten, D. Zhu, S. Wei, Femtosecond atomic-scale X-ray photon correlation spectroscopy enables direct probing of relaxation dynamics in fragile liquid, Dec 2023. arXiv:2312.08613 [cond-mat]
31. J. Miao, T. Ishikawa, I.K. Robinson, M.M. Murnane, Beyond crystallography: diffractive imaging using coherent X-ray light sources. Science **348**(6234), 530–535 (2015)
32. B. Abbey, K.A. Nugent, G.J. Williams, J.N. Clark, A.G. Peele, M.A. Pfeifer, M. de Jonge, I. McNulty, Keyhole coherent diffractive imaging. Nat. Phys. **4**(5), 394–398 (2008)
33. J.R. Fienup, Reconstruction of an object from the modulus of its Fourier transform. Opt. Lett. **3**(1), 27–29 (1978)
34. J. Fan, J. Zhang, Z. Liu, Coherent diffraction imaging of cells at advanced X-ray light sources. TrAC Trends Anal. Chem. **171**, 117492 (2024)
35. T. Wang, D. Zhu, W. Benny, C. Graves, S. Schaffert, T. Rander, L. Müller, B. Vodungbo, C. Baumier, D.P. Bernstein, B. Bräuer, V. Cros, S. De Jong, R. Delaunay, A. Fognini, R. Kukreja, S. Lee, V. López-Flores, J. Mohanty, B. Pfau, H. Popescu, M. Sacchi, A.B. Sardinha, F. Sirotti, P. Zeitoun, M. Messerschmidt, J.J. Turner, W.F. Schlotter, O. Hellwig, R. Mattana, N. Jaouen, F. Fortuna, Y. Acremann, C. Gutt, H.A. Dürr, E. Beaurepaire, C. Boeglin, S. Eisebitt, G. Grübel, J. Lüning, J. Stöhr, A.O. Scherz, Femtosecond single-shot imaging of nanoscale ferromagnetic order in Co/Pd multilayers using resonant X-ray holography. Phys. Rev. Lett. **108**(26), 267403 (2012)
36. F. Pfeiffer, X-ray ptychography. Nat. Photonics **12**(1), 9–17 (2018)
37. K.A. Nugent, Coherent methods in the X-ray sciences. Adv. Phys. **59**(1), 1–99 (2010)
38. H.N. Chapman, K.A. Nugent, Coherent lensless X-ray imaging. Nat. Photonics **4**(12), 833–839 (2010)
39. F. Livet, Diffraction with a coherent X-ray beam: dynamics and imaging. Acta Crystallogr. Sect. A Found. Crystallogr. **63**(2), 87–107 (2007)
40. L. Wiegart, O. Chubar, A. Fluerasu, M.S. Rakitin, X-ray optical simulations supporting advanced commissioning of the coherent hard X-ray beamline at NSLS-II, in *Proceedings of SPIE 10388, Advances in Computational Methods for X-Ray Optics IV*, vol. 103880N, Aug 2017, p. 103880N–1 (2017). ISBN: 9781510612334
41. D. Attwood, A. Sakdinawat, *X-Rays and Extreme Ultraviolet Radiation: Principles and Applications*, 2nd edn. (Cambridge University Press, 2017)
42. A.R. Sandy, Q. Zhang, L.B. Lurio, Hard X-ray photon correlation spectroscopy methods for materials studies. Annu. Rev. Mater. Res. **48** (2018)
43. M. Sutton, A review of X-ray intensity fluctuation spectroscopy. C. R. Phys. **9**(5–6), 657–667 (2008)
44. J. Li, M. Madhavi, S. Jeppson, L. Zhong, E.M. Dufresne, B. Aitken, S. Sen, R. Kukreja, Observation of collective molecular dynamics in a chalcogenide glass: results from X-ray photon correlation spectroscopy. J. Phys. Chem. B **126**(28), 5320–5325 (2022)
45. G. Pintori, G. Baldi, B. Ruta, G. Monaco, Relaxation dynamics induced in glasses by absorption of hard X-ray photons. Phys. Rev. B **99**(22), 224206 (2019)

46. M. Sutton, J.R.M. Lhermitte, F. Ehrburger-Dolle, F. Livet, High resolution strain measurements in highly disordered materials. Phys. Rev. Res. **3**(1), 013119 (2021)
47. J. Li, Capturing fluctuation dynamics in functional materials using X-ray photon correlation spectroscopy. Ph.D. thesis, University of California Davis, 2021
48. B.J. Berne, *Dynamic Light Scattering: With Applications to Chemistry, Biology, and Physics*, Dover edn. (Dover Publications, Mineola, NY, 2000, 1976)
49. P. Falus, L.B. Lurio, S.G.J. Mochrie, Optimizing the signal-to-noise ratio for X-ray photon correlation spectroscopy. J. Synchrotron Radiat. **13**(3),253–259 (2006)
50. G. Grübel, D.L. Abernathy, D.O. Riese, W.L. Vos, G.H. Wegdam, Dynamics of dense, charge-stabilized suspensions of colloidal silica studied by correlation spectroscopy with coherent X-rays. J. Appl. Crystallogr. **33**(3 Part 1), 424–427 (2000)
51. L.B. Lurio, D. Lumma, A.R. Sandy, M.A. Borthwick, P. Falus, S.G.J. Mochrie, J.F. Pelletier, M. Sutton, L. Regan, A. Malik, G.B. Stephenson, Absence of scaling for the intermediate scattering function of a hard-sphere suspension: static and dynamic X-ray scattering from concentrated polystyrene latex spheres. Phys. Rev. Lett. **84**, 785–788 (2000)
52. L. Cipelletti, S. Manley, R.C. Ball, D.A. Weitz, Universal aging features in the restructuring of fractal colloidal gels. Phys. Rev. Lett. **84**(10), 2275–2278 (2000). ISBN: 0031-9007
53. Z. Evenson, B. Ruta, S. Hechler, M. Stolpe, E. Pineda, I. Gallino, R. Busch, X-ray photon correlation spectroscopy reveals intermittent aging dynamics in a metallic glass. Phys. Rev. Lett. **115**(17), 175701 (2015)
54. C. Sanborn, K.F. Ludwig, M.C. Rogers, M. Sutton, Direct measurement of microstructural avalanches during the martensitic transition of cobalt using coherent X-ray scattering. Phys. Rev. Lett. **107**(1), 1–4 (2011)
55. J.-P. Bouchaud, E. Pitard, Anomalous dynamical light scattering in soft glassy gels. Eur. Phys. J. E **6**(3), 231–236 (2001)
56. E. Sanz, C. Valeriani, E. Zaccarelli, W.C.K. Poon, M.E. Cates, P.N. Pusey, Avalanches mediate crystallization in a hard-sphere glass. Proc. Natl. Acad. Sci. **111**(1), 75–80 (2014)
57. A. Duri, L. Cipelletti, Length scale dependence of dynamical heterogeneity in a colloidal fractal gel. Europhys. Lett. (EPL) **76**(5), 972–978 (2006)
58. A.S. Everhardt, S. Matzen, N. Domingo, G. Catalan, B. Noheda, Ferroelectric domain structures in low-strain $BaTiO_3$. Adv. Electron. Mater. **2**, 1500214 (2016)
59. A.S. Everhardt, S. Damerio, J.A. Zorn, S. Zhou, N. Domingo, G. Catalan, E.K.H. Salje, L.Q. Chen, B. Noheda, Periodicity-doubling cascades: direct observation in ferroelastic materials. Phys. Rev. Lett. **123**, 087603 (2019)
60. P. Zubko, N. Stucki, C. Lichtensteiger, J.M. Triscone, X-ray diffraction studies of 180° ferroelectric domains in $PbTiO_3/SrTiO_3$ superlattices under an applied electric field. Phys. Rev. Lett. **104**, 187601 (2010)
61. P. Zubko, N. Jecklin, N. Stucki, C. Lichtensteiger, G. Rispens, J.M. Triscone, Ferroelectric domains in $PbTiO_3/SrTiO_3$ superlattices. Ferroelectrics **433**(1), 127–137 (2012)
62. L. He, D. Vanderbilt, First-principles study of oxygen-vacancy pinning of domain walls in $PbTiO_3$. Phys. Rev. B **68**(13), 134103 (2003)
63. J. Schlappa, C. Schüßler-Langeheine, C.F. Chang, H. Ott, A. Tanaka, Z. Hu, M.W. Haverkort, E. Schierle, E. Weschke, G. Kaindl, L.H. Tjeng, Direct observation of t_{2g} orbital ordering in magnetite. Phys. Rev. Lett. **100**(2), 1–4 (2008)
64. D.J. Huang, H.-J. Lin, J. Okamoto, K.S. Chao, H.-T. Jeng, G.Y. Guo, C.-H. Hsu, C.-M. Huang, D.C. Ling, W.B. Wu, C.S. Yang, C.T. Chen, Charge-orbital ordering and Verwey transition in magnetite measured by resonant soft X-ray scattering. Phys. Rev. Lett. **96**(9), 1–4 (2006)
65. M.S. Senn, J.P. Wright, J.P. Attfield, Charge order and three-site distortions in the Verwey structure of magnetite. Nature **481**(7380), 173–176 (2012)

Resonant Hard X-Ray Scattering Probes of Electronic and Magnetic Order: A Practical Guide

Derek J. Meyers, John W. Freeland, and Philip J. Ryan

Abstract Quantum materials host a large variety of emergent phenomena including magnetism, superconductivity, and many more (Tokura and Nagaosa in Science 288:462, 2000 [1]; Keimer and Moore in Nat Phys 13:1045, 2017 [2]). Understanding these unique and often interconnected phases requires probes for the charge, orbital, and spin configurations in these systems. Resonant elastic X-ray scattering (REXS) has emerged as a powerful tool to extract these details which are essential to building a clear picture of the underlying physics as presented in detail in several excellent reviews (Subías et al. in J Phys Conf Ser 190, 2009 [3]; Beale in Eur Phys J Spec Top 208:89, 2012 [4]; Vettier in Eur Phys J Spec Top 208:3, 2012 [5]). In order to make this approach more accessible to the non-expert, the goal of this chapter is to use examples of REXS as the basis for a more practical tutorial about common pitfalls and how to harness the power of this technique to explore electronic and magnetic order in quantum materials. While we focus on hard X-rays we note that soft X-ray resonant scattering is also highly useful and versatile (Fink et al. in Rep Prog Phys 76, 2013 [6]). This chapter does not consider inelastic interactions between the samples and X-ray photons, for example as in inelastic X-ray scattering (RIXS) (Ament et al. in Rev Mod Phys 83:705, 2011 [7]). The interested reader is encouraged to explore two excellent reviews focused on those techniques (Fink et al. in Rep Prog Phys 76, 2013 [6]; Ament et al. in Rev Mod Phys 83:705, 2011 [7]).

Keywords Resonant X-ray scattering · Charge ordering · Orbital ordering · Magnetic ordering · Anomalous dispersion correction · Polarized X-ray optics

D. J. Meyers (✉)
Department of Physics, Oklahoma State University, Stillwater, OK, USA
e-mail: derek.meyers@okstate.edu

J. W. Freeland · P. J. Ryan
X-Ray Science Division, Argonne National Laboratory, Lemont, IL, USA
e-mail: freeland@anl.gov

P. J. Ryan
e-mail: pryan@anl.gov

D. Sando et al. (eds.), *X-Ray Scattering Techniques for Epitaxial Oxide Thin Films*,
https://doi.org/10.1007/978-981-96-5945-6_5

1 A Brief Introduction of the Fundamental Theoretical and Experimental Principles

1.1 Energy Dispersive X-Ray Scattering

The scattering of X-rays by matter is a complex process which allows investigation of numerous physical properties of materials. Laboratory-scale X-ray scattering is often used to determine the material composition and, when combined with phase retrieval methods, the position of atoms inside ordered (i.e. crystalline), amorphous and glassy materials. However, a wealth of information can be gleaned beyond simple atomic structure using X-ray scattering.

Fundamentally, the interaction of an incident photon with charged particles can be understood through the quantized magnetic and electric fields of the photon with the electric charge and magnetic moment of the particle, typically an electron. The electromagnetic vector potential, $\vec{A}$, of the X-ray radiation can be used to describe several of the important phenomena arising from this interaction. In its quantum mechanical form, the vector potential can be expressed in terms of photon creation and annihilation operators. The interaction of the photon with the particle leads to several linear and quadratic terms in the interaction Hamiltonian that can create or annihilate photons, or combinations thereof [8]. The linear term corresponds to the absorption or emission of a photon and does not contribute to the scattering cross-section. The quadratic terms in $\vec{A}$ lead to several processes which are collectively referred to as scattering terms, except in the cases of absorption (annihilation) or spontaneous emission (creation) of two photons simultaneously, which have very low probabilities and are typically ignored.

A full treatment of the photon-electron interaction can be found in review articles and books [6, 9–11]. Here, we will broadly discuss the interaction and the important terms that lead to the phenomena discussed in this chapter. The non-relativistic interaction Hamiltonian, H_{in}, between an arbitrary charged particle and a photon can be written as:

$$H_{in} = -\frac{q}{m}\vec{p} \cdot \vec{A} + -\frac{q^2}{2m}\vec{A} \cdot \vec{A} - g\frac{q}{2m}\vec{s} \cdot \vec{B} \tag{1}$$

Here q is the charge of the particle, m is the mass of the particle, $\vec{p}$ is the free momentum of the particle, $\vec{A}$ is the magnetic vector potential of the photon, g is the gyromagnetic ratio of the particle, $\vec{s}$ is the spin of the particle, and $\vec{B}$ is the oscillatory magnetic field of the photon. We ignore the terms for the particle and photon non-interacting energies as they will not affect the scattering results. Furthermore, the direct coupling of the particle magnetic moment $\vec{s}$ with the photon is a relativistic term with a factor of $\frac{\hbar\omega}{mc^2}$ making it typically very small.[1] The cross-section for each

[1] In the case of very high energy X-ray scattering (> 60 keV) this term can be appreciable. However, the large penetration depth of these X-rays (typically ~ mm) limits it to larger sample volumes that may make neutron magnetic scattering more feasible, depending on the elements present.

interaction goes as the square of these terms, so the $\frac{q}{m}$ pre-factor strongly suggests the interaction will be dominated by photon-particle interactions for electrons. Photon-nucleus interactions are much less important, due to the limited potential total electric charges compared to the more than three orders of magnitude difference in mass.

To find the photon scattering cross-section for electrons, one applies Fermi's Golden Rule with H_{in} from Eq. 1 as a perturbation. The transition operator T_i for the interaction Hamiltonian to second order is given by:

$$T_i \approx H_{in} + H_{in} G_0(E_i) H_{in} \tag{2}$$

Here $G_0(E_i)$ is the resolvent of H_{in} at energy E_i, with H_{in} taken as a perturbation of the transition operator. This is defined as:

$$G(E_i) = \lim_{\epsilon \to 0^+} \frac{1}{E_i - H_0 + i\epsilon} \tag{3}$$

Putting this together and using $q = e$, we obtain the following expression for the transition operator, where we neglect terms of higher order than $\frac{e^2}{m^2}$:

$$T_i \approx -\frac{e}{m}\vec{p} \cdot \vec{A} - \frac{e^2}{2m}\vec{A} \cdot \vec{A} - \frac{e^2}{m^2}\vec{p} \cdot \vec{A} G_0(E_i) \vec{p} \cdot \vec{A} \tag{4}$$

Equation 4 has terms that are linear and quadratic in the magnetic vector potential $\vec{A}$ (or field $\vec{B}$). This quantized vector potential can be expressed as a linear combination of raising and lowering operators, annihilating or creating a photon:

$$\vec{A} = \sqrt{\frac{2\pi \hbar c}{V}} \sum \frac{1}{\sqrt{k}} \hat{e}_i \left(a e^{iQ \cdot r} + a^\dagger e^{-iQ \cdot r} \right) \tag{5}$$

Here $\hat{e}_i$ is the photon polarization direction, a and $a^\dagger$ are the raising and lowering operators, Q is wavevector of the created or annihilated photon, and is now the created or annihilated photon wave vector, and r is the position of the scattering atom.

Linear terms in $\vec{A}$ lead to several specific processes broadly grouped into either photon emission or photon absorption. Radiative deexcitation occurs when an electron drops from a higher energy orbital to a lower one, creating a distinct energy photon. Electron trapping occurs when electrons from non-distinct continuum states drop to distinct states releasing photons with a range of energies. Finally, Bremsstrahlung radiation occurs when high energy continuum electrons lose energy in the form of X-ray photons but do not become trapped. Conversely, resonant absorption occurs when a photon gives an electron a distinct energy to transition to a higher energy state. Photoionization occurs when the electron is excited into a continuum of high energy states with a range of energies. Finally, inverse Bremsstrahlung occurs when a photon raises a continuum electron to higher energy without fully stripping it from the atom. These processes underlie numerous spectroscopic techniques and are all important in their own right, but for this review we are interested in the quadratic terms in $\vec{A}$ which lead to elastic scattering events.

The first quadratic term in $\vec{A}$ from Eq. 4 applied to the initial and final states of the system depends on the incoming and outgoing photon polarization and the momentum transfer vector $\vec{Q} = \overrightarrow{q_f} - \overrightarrow{q_i}$ for the incoming and outgoing photon momentums, $\overrightarrow{q_f}$ and $\overrightarrow{q_i}$, respectively. If the energy of the incoming and outgoing photons is not changed then the first quadratic term in Eq. 4 describes elastic scattering which is termed Thompson scattering. This encompasses basic X-ray scattering and reflectivity and has large enough cross-sections to utilize laboratory scale X-ray sources.

The second term quadratic in $\vec{A}$, $\vec{p} \cdot \vec{A} G_0(E_i) \vec{p} \cdot \vec{A}$, is highly energy dependent owing to the resolvent G_0 and is typically referred to as the resonant scattering or energy dispersive term. Here an intermediate state with no photons occurs as the first term is applied to the initial state. After the second term is applied the denominator consists of $E_f - E_i - h\omega$. Resonance refers to the condition where the difference between the electron energy in the initial and intermediate states matches the incident photon energy $h\omega_i$, with the denominator approaching zero the cross-section rapidly increases approaching and decreases away from the resonant condition. The available initial and intermediate electron states are determined by the atoms, their bonding, local environment, and the Fermi level of the system. Thus, the energy dependence of this term makes it sensitive to the electronic structure of the system, in addition to the Q dependence giving additional information not available in Thompson scattering experiments. Furthermore, the polarization dependence, resulting from the form of $\vec{A}$, leads to additional important consequences.

The total transition probability using time dependent perturbation theory (i.e. applying Fermi's Golden Rule), has the following form:

$$w_{i \to f} = \frac{2\pi}{\hbar} \left| \langle \phi_f | T_i | \phi_i \rangle^2 \right| \rho_f \tag{6}$$

Here w is the transition probability, ϕ_f and ϕ_i are the final and initial states, respectively, and ρ_f is the density of final states. Taking Eqs. 4 and 5 we can derive the Kramers–Heisenberg differential cross-section for elastic scattering:

$$\frac{d\sigma}{d\Omega} = \frac{e^2}{mc^2} \frac{w_{q_f}}{w_{q_i}} \left| \hat{e}_f \cdot \hat{e}_i \langle \phi_f | e^{-iQ \cdot r} | \phi_i \rangle + \sum_n \frac{\langle \phi_f | \hat{e}_f \cdot \vec{p} e^{iq_f \cdot r} | \phi_n \rangle \langle \phi_n | \hat{e}_i \cdot \vec{p} e^{-iq_i \cdot r} | \phi_i \rangle}{E_i - E_n + h\omega + \frac{i\Gamma}{2}} \right|^2 \tag{7}$$

The differential cross-section is the probability that a photon will be scattered into solid angle $d\Omega$. Here $\vec{Q}$ is simply $\overrightarrow{q_f} - \overrightarrow{q_i}$, $|\phi_n\rangle$ is the intermediate state, and Γ is the lifetime of the intermediate state. The first term is the familiar Thompson scattering cross-section. The second term is the resonant scattering cross-section.[2] For this chapter, we are only interested in elastic resonant scattering, so $|\phi_f\rangle = |\phi_i\rangle$, readers interested in the inelastic scattering, where $|\phi_f\rangle \neq |\phi_i\rangle$ are directed to the excellent review by Ament et al. [7]. The differential cross-section can be written in terms of what is called the scattering length f:

$$\frac{d\sigma}{d\Omega} = |f|^2 \tag{8}$$

This scattering length can then be broken down into the following form:

$$f = f^T + \Delta f' + i\Delta f'' \tag{9}$$

Here f^T is the Thompson scattering length, and $\Delta f'$ ($\Delta f''$) is the real (imaginary) part of the anomalous dispersion correction Δf to the scattering.[3] $\Delta f'$ and $\Delta f''$ are what is measured for both resonant X-ray scattering and resonant X-ray absorption spectroscopy, respectively. f^T is a function only of the scattering vector $\vec{Q}$ and the X-ray polarizations, $\hat{e}_i$ and $\hat{e}_f$, whereas Δf depends on the energy of the incident photons, $h\omega$, the X-ray polarizations, $\hat{e}_i$ and $\hat{e}_f$, and the incident and scattered X-ray wave vectors, q_f and q_i. Thus, the anomalous dispersion correction can reveal much more detailed information for a system than simple Thompson scattering.

Furthermore, these terms can be simplified using the dipole approximation, where we ignore the spatial variation of $\vec{A}$. This then assumes that $e^{iQ\cdot r} \cong 1$. This only allows elastic scattering and has no dependence on the momentum. For the first term there is then only polarization dependence. However, the second term retains both its polarization and energy dependence. This simplification then allows for analytical results. The appropriateness of this approximation can be understood most easily by picturing the Fourier transform of the different electronic structures that contributes to each term. For Thompson scattering, the outer electron shells contribute the vast majority to the scattering cross-section, these are the most broadly distributed orbitals in real space. The Fourier transform of these voluminous real space orbitals, which fill the crystal, result in sharp reflections in reciprocal space, making the approximation less useful. However, cross-sections for energy dispersive terms are dominated by the core electron levels, which are very localized near the nucleus, when converted to reciprocal space this then gives broad $\vec{Q}$-dependence allowing for the assumption $e^{iQ\cdot r} \cong 1$. Use of the dipole approximation and the transition probability per unit time allows the cross-section to be written in a very compact form after several intermediate steps [10]:

[2] Note there is another ignored term identical to the resonant term except with $-h\omega_i$ in the denominator. However, the denominator never approaches 0 and thus is exceedingly weak at X-ray energies.

[3] There is another term, f^M the magnetic scattering length, that comes from terms we dropped from the derivation earlier due to its weak cross-section at typical X-ray energies.

$$\frac{d\sigma}{d\Omega} = \frac{4\pi}{q_i} Im[\Delta f] = \frac{4\pi}{q_i} \Delta f'' \tag{10}$$

This is known as the optical theorem, and it allows $\Delta f''$ to be found using only X-ray absorption spectroscopy. Importantly, Kramers–Kronig [12] relations then allow extraction of $\Delta f'$ meaning the entire form of the rank-2 tensor Δf can be deciphered. In practice, finding Δf from measurements of X-ray absorption spectroscopy ($\Delta f''$) requires studies with polarization aligned along each crystal principal axis along with several linear combinations thereof to fully extract $\Delta f'$. Experimentally, this can be time consuming, and it is easier to do a direct measurement of $\Delta f'$ as discussed later.

Experimentally, resonant diffraction is typically analyzed using a simplified model of a free magnetic ion with spherical symmetry, which allows a straightforward analytical solution for Δf [9]:

$$\Delta f = f^0 \hat{e}_f \cdot \hat{e}_i - i f^1 (\hat{e}_f \times \hat{e}_i) \cdot \hat{s} + f^2 (\hat{e}_f \cdot \hat{s})(\hat{e}_f \cdot \hat{s}) \tag{11}$$

Here f^j (j = 0, 1, 2) are the energy dependent resonance factors for the allowed electronic transitions of the atom. These three terms correspond to three distinct types of resonant scattering. The first term does not involve the atomic spin and is thus called non-magnetic resonant scattering, this encompasses charge and orbital resonant scattering, discussed below. The second term is linearly dependent on the spin and is thus called resonant magnetic scattering. The third term is quadratic in the spin and is also termed resonant magnetic scattering. Typically, the f^2 term is small compared to f^1 and is often ignored.

In real systems the spherical symmetry used for Eq. 11 is not realistic. A full treatment without any symmetry considerations, ignoring the higher order f^2 terms, is given by $\Delta f = \hat{e}' \cdot \Delta f^* \cdot \hat{e}$ with the tensor:

$$\Delta f^* = \begin{pmatrix} f_{xx} & f_{xy} & f_{xz} \\ f_{yx} & f_{yy} & f_{yz} \\ f_{zx} & f_{zy} & f_{zz} \end{pmatrix} \tag{12}$$

The higher order terms can be important in certain situations and can be found as well, but we ignore them here [13]. This general expression can then be simplified by any symmetries which are present. In the case of the free magnetic ion, where spherical symmetry is only broken by the spin, one finds:

$$\Delta f^* = f^0 \begin{pmatrix} 1 & 0 & 0 \\ 0 & 1 & 0 \\ 0 & 0 & 1 \end{pmatrix} + f^1 \begin{pmatrix} 0 & -z & y \\ z & 0 & -x \\ -y & x & 0 \end{pmatrix} \tag{13}$$

Which is equivalent to Eq. 11 above with the f^2 term ignored. Other symmetry breakings, from the lattice, orbitals, etc., can then further complicate the final expression, but typically Eq. 13 is close enough for prediction of experimental data.

The polarization of X-rays from synchrotron sources is often horizontally polarized, denoted $\hat{e}_\sigma$, and can be flipped to vertically polarized, denoted $\hat{e}_\pi$. For most experiments then we can write the total Δf in terms of a 2 × 2 scattering matrix:

$$S = \begin{pmatrix} \hat{e}'_\sigma \cdot \Delta f^* \cdot \hat{e}_\sigma & \hat{e}'_\sigma \cdot \Delta f^* \cdot \hat{e}_\pi \\ \hat{e}'_\pi \cdot \Delta f^* \cdot \hat{e}_\sigma & \hat{e}'_\pi \cdot \Delta f^* \cdot \hat{e}_\pi \end{pmatrix} \tag{14}$$

This matrix is then summed over the lattice sites to find the total scattering amplitude. When we consider a free magnetic atom with $\hat{s} = (x\ y\ z)$ this can be written as [9]:

$$S = f^0 \begin{pmatrix} 1 & 0 \\ 0 & \cos(2\theta) \end{pmatrix} - if^1 \begin{pmatrix} 0 & x\cos(\theta) + z\sin(\theta) \\ z\sin(\theta) - x\cos(\theta) & -y\sin(2\theta) \end{pmatrix} \tag{15}$$

where we again ignore the f^2 term. Here, the expression assumes vertical scattering geometry, with θ being the scattering angle from Fig. 2. The first term corresponds to Thompson scattering and has no polarization dependence other than the geometrical factor, but the magnetic term has distinct cross channel terms. If we consider a spin oriented in the plane of the sample surface in bisecting geometry, with the azimuthal angle φ, i.e. $\hat{s} = (\cos\varphi\ \sin\varphi\ 0)$, describing the rotation of the sample about the normal as depicted in Fig. 2, the magnetic scattering term is:

$$S_M = -if^1 \begin{pmatrix} 0 & \cos\Psi\cos(\theta) \\ -\cos\Psi\cos(\theta) & -\sin\Psi\sin(2\theta) \end{pmatrix} \tag{16}$$

This leads to distinct dependence of the scattering intensity on all polarization channels ($\pi'\pi$, $\pi'\sigma$, $\sigma'\pi$, and $\sigma'\sigma$) and the azimuthal angle. In a typical magnetic scattering experiment, the cross channel is used to map the azimuthal dependence of the intensity, allowing extraction of $\hat{s}$. Similar experiments for moments along the c-axis or any arbitrary direction can be done, albeit with different forms for S_M. Unfortunately, in the case of ferromagnetic ordering the magnetic reflection and Thompson scattering occur at the same Q, and f^1 tends to be on the order of 10^6 weaker. With perfect polarization filtering this would not be an issue, but real polarization analysis crystals typically only have ~ 10^{-2} filtering efficiency. While mostly applicable to antiferromagnetic ordering, where the larger magnetic lattice gives peaks which have no charge contribution, using both polarization discrimination with the sample diffraction (2θ) around 90°,[4] combined with post sample analyzers and reasonable resonance effects make ferromagnetic order observable [14]. This requires that: (i) a scattering angle of ~ 90° for a relevant peak at the resonant edge energy, and (ii) suitable polarization analysis crystal (or grating for soft X-rays) for the edge energy. The general limitation for magnetic scattering, particularly in the case of thin films, is the magnitude of the cross-section required to have a measurable signal when

[4] Scattering angles at exactly 90° prevent charge scattering by virtue of the $\hat{e}_f \cdot \hat{e}_i$ factor when the polarization is parallel to the scattering plane.

using an analyzer crystal. The resonant cross-section is dependent on the number of unoccupied density of states of the final state, this makes the L edges (p–d) transition metals at soft energies and the rare earth systems at higher energies very attractive for ferromagnetic order studies.

The equations above were derived for spherical symmetry. When ignoring the quadratic term in $\hat{s}$ this form still works for cubic symmetry, but is no longer valid for lower symmetries, eventually returning to the general form in Eq. 12. The deviation from the cubic approximation for S usually tracks with small deviations from cubic. However, the polarization channel distinction may lose validity making magnetic and charge signals more difficult to distinguish.

Resonant scattering has a very weak Q-dependence in the dipole approximation. If the ordering is not known beforehand finding a particular vector of the ordering is not trivial for single crystal systems. High energy (> 100 keV) X-ray diffraction, powder scattering experiments, either neutron or X-ray are ideal for survey studies looking for unknown ordering behavior.

Finally, the energy dependence of the resonant scattering terms, encoded in the f terms, is a function of the local crystal environment, valence, and bonding and thus contains detailed, but complex information. Especially in the case of highly correlated materials, theoretical calculations are needed to decipher the energy profile.

As an example, Fig. 1 shows energy dispersive terms as function of energy for an iridate material at photon energies near the L_3 absorption edge. The imaginary term data comes from X-ray absorption spectroscopy and the real term data is constructed from differential Kramers–Kronig analysis [15] of published data [16]. The top panel shows the charge scattering factor and the bottom panel the magnetic scattering factor, i.e. f^0 and f^1 respectively.

1.2 Experimental Apparatus

Measurement of resonant phenomena requires the ability to tune the energy of the incident photons to match the electronic transitions of the sample, i.e. where the f terms become large. Thus, these measurements are limited to synchrotron and free electron laser (FEL) facilities in practice. Both of these types of facilities generate high brilliance X-ray beams through manipulation of electrons. In the case of synchrotrons, an electron storage ring houses a high energy (GeV) electron beam which is maintained by constant or periodic addition from a linear electron accelerator. X-ray FELs generate a linear electron beam which travels through an extremely long undulator. The interaction of the electron beam and the generated X-rays leads to microbunching for the electrons which generates extremely coherent and time-resolved X-ray packets. This process limits these devices to a single X-ray beam operation at any given time, but the coherence and time resolution gives them a unique purpose for specialized experiments. In both types of facilities, the electrons travel through insertion devices or bending magnets which periodically perturbs the electron trajectory generating synchrotron X-rays. The energy of these X-rays is

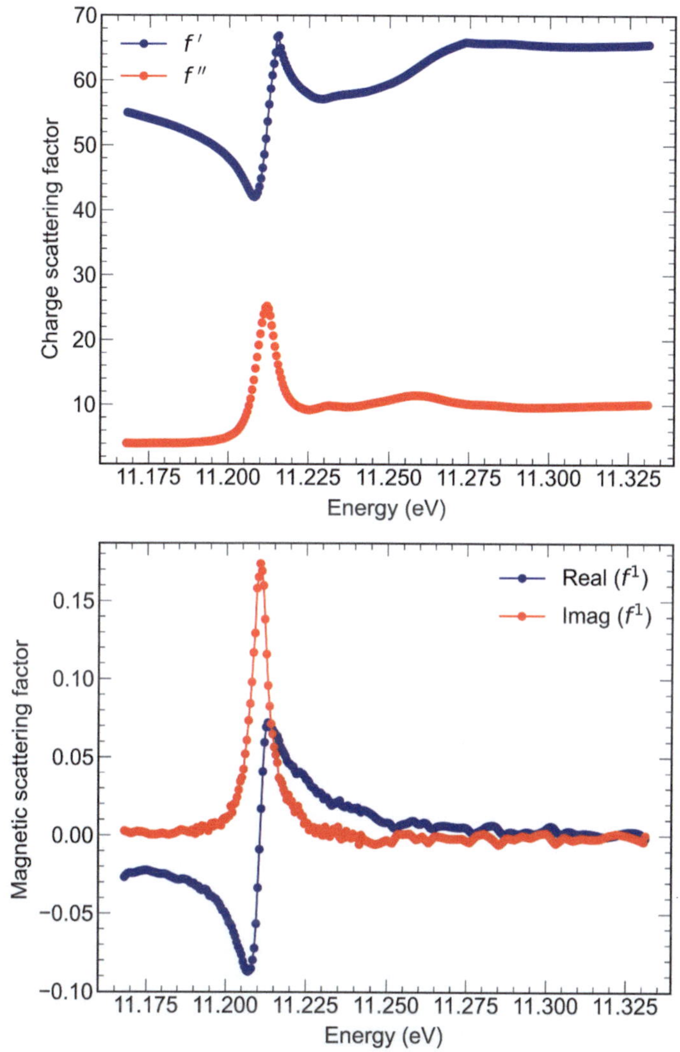

Fig. 1 Real and imaginary scattering lengths for **a** charge and **b** magnetic processes. Data shown for iridium based material at the photon energies near the L_3 resonance

characteristic of the insertion device (ID) parameters and typically covers a broad energy spectrum. The IDs are capable of adjusting the range of energies produced on-demand. The broad energy beam is then put through a series of optics which remove all but the desired X-ray energies, with the energy resolution highly variable based upon the optics and technique. These optics vary based on energy range and resolution required for the different scattering techniques.

As discussed in Sect. 1.1, resonant scattering techniques are sensitive to the energy, scattering vector, $\vec{Q}$, and polarization of the incident and scattered X-rays. Due to

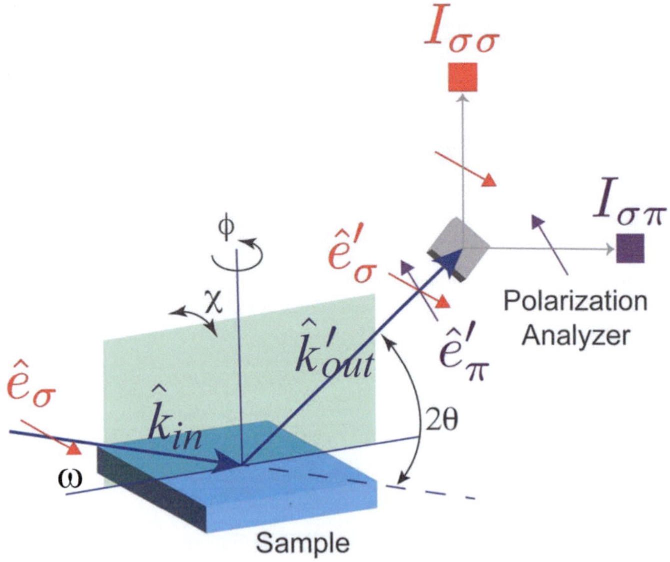

Fig. 2 Scattering geometry and polarization analyzer necessary for most resonant scattering experiments. The scattering plane defined by the incoming and outgoing X-ray beam is highlighted in green

all of these factors the experimental setups can vary quite widely depending on the specific measurements a beamline is designed to perform.

Typically, the incident energy dependence and polarization are controlled by the beamline optics which are well separated from the actual experimental apparatus. Control of the scattering vector and analysis of the diffracted beam polarization and energy are beamline and sample dependent, a typical vertical scattering geometry and instrument is shown in Fig. 2. The incoming beam polarization can often be switched with phase plates from σ to π depending on the beamline optics, but typically at some cost of flux [17]. For most experimental setups, there is a minimum of 4 circles on the goniometer to position the sample and set the $\vec{Q}$ to the necessary scattering conditions. These are the sample incident X-ray angle, ω, the angle about the scattering plane, χ, the sample rotation about its surface normal, φ, and the detector angle relative to the incoming X-rays, 2θ. More complex 6-circle diffractometers can have additional circles, which, for example, can change the η and χ motions relative to the scattering plane, μ, or the angle of the detector arm relative to the incident beam in the horizontal plane, ν. The azimuthal angle, Ψ, is dependent on the relative orientation of the surface normal and scattering vector and thus depends on multiple sample angles. In the special case of a parallel scattering vector and surface normal $\Psi = \phi$. It should be noted that different conventions for the angle designations exist, going from 4 to 6 circle scattering modes, different instruments, different motor stacking schemes, etc. Furthermore, radically different instruments exist which have unique advantages and disadvantages, for example Kappa versus Eulerian geometry diffractometers.

In the more typical Eulerian setup in Fig. 2, the X-ray beam exits a Be window from the evacuated beampipe, travels through the air to the sample, then from the sample to the diffraction arm tube through another Be window. Because the incident beam propagates through air to reach the sample, the X-ray energies that can be used are limited to those where air scattering is minimal, above ~ 3 keV. This is near the transition from the tender (~ 1.5–5 keV) to the hard X-ray regime (5–200 keV). These instruments tend to be very customizable and allow for numerous scattering geometries and can sample a large area of Q-space, due to both the ability to measure in air and the higher X-ray energies and thus higher $\vec{Q}$ values.

In the case of soft (200–1500 eV) and, to a lesser extent, tender X-ray scattering experiments, typically air and Be scattering are very strong and the beam must be kept in vacuum or low pressure He gas environments, respectively. Due to this the number of circles can be more limited though many modern chambers incorporate four or more circles. In addition, the lower X-ray energy means the probed Q-space is much smaller (< 1 Brillouin zone typically for soft X-rays).

Polarization analysis of the scattering beam can be highly insightful by allowing clear distinction between charge (i.e. lattice, orbital, or charge ordering) and magnetic scattering as described in Sect. 1.1. X-rays produced in undulators are naturally horizontally polarized. Thus, many experimental setups are optimized to utilize this natural polarization to maintain high levels of flux. However, when vertically polarized or circularly polarized X-rays are required X-ray optics can be used, with some flux cost. Also, scattering geometry can be switched from vertical to horizontal to change incident polarization from σ to π.

Focusing on the common hard X-ray experiment, horizontally polarized X-rays (σ) are incident upon the sample in a vertical scattering geometry (incident and outgoing X-rays form a 2D plane perpendicular to the floor). For charge scattering, the scattering beam maintains the same polarization (σ) while for magnetic scattering with the magnetic moment in the diffraction plane, the photon angular momentum flips the electron spin orientation and through conservation of angular momentum has its polarization switched to vertically polarized (π). This beam is then incident upon an analyzer crystal diffracting perpendicular to the diffraction plane of the sample which has a reflection with a Bragg angle as close to $\frac{\pi}{2}$ as possible by selecting a crystal system and allowed plane of diffraction to match the chosen incident energy. In this geometry the analyzer selects only π polarized X-rays and σ polarized X-rays are optimally suppressed (by a factor of $\cos(2\theta)$). Vertically polarized X-rays will not be affected as $\hat{e}_f \cdot \hat{e}_i = 1$ for the analyzer crystal scattering only affected by the mosaicity of the analyzer and the cross-section of the particular reflection.

2 Tutorial Style Instructions on Obtaining Practical Data

2.1 Sample Requirements

Before attempting a resonant scattering experiment several requirements must be addressed regarding the sample. These will vary greatly considering the measurements needed and the beamline setup. This section will serve as a general guide for the types of consideration that must be made, but more specialized experiments may have further requirements.

Resonant scattering can be performed utilizing any transition within the sample's atomic composition. However, it is important that the phenomena being investigated, i.e. magnetic, orbital, or charge ordering, involves the atoms whose resonance is being utilized. This will typically limit the potential resonances that can be used. Furthermore, the electron needs to be excited to an active orbital, i.e. one close enough to the Fermi level to be affected by the phenomena being investigated. There can be exceptions due to accompanying structural symmetry changes, which can indirectly probe the ordering. Establishing the resonance of interest will typically limit the experiment to the hard, tender, or soft X-ray regimes. The available scattering beamlines which cover this energy range can then be chosen. Related to this, the energy range chosen can also limit the range of reciprocal space that can be accessed so it is equally important to make sure that the relevant scattering conditions can be reached. Beyond direct calculation, there are numerous tools available for this task, including powder diffraction simulators in freely available software such as Vesta or CrystalDiffract [18, 19].

While these calculators are for powder diffraction, they are a very useful tool for highlighting allowed peaks which are essential for orienting the sample as an important first step. Note, however, that peaks related to magnetic or electronic order are not captured by these calculators and that aspect might require a separate calculation of the structure factor. The first step in an experiment will be aligning the reciprocal space of the lattice. Once a sample of known orientation is mounted in the experimental apparatus, alignment is accomplished using a minimum of two structural reflections. For hard X-rays this is typically trivial due to the large $\vec{Q}$ range accessible. However, for soft X-ray experiments, especially at the lower energy extremes, the scattering vector $\vec{Q}$ is often limited to less than a single Brillouin zone. In this case if the surface is of sufficient quality the specular reflection can be used to align the η and χ offsets, starting at very low 2θ to aid in finding the reflection, and then gradually increasing 2θ while aligning η and χ at each step until relatively high angle (~ 90°) is achieved to increase precision. With this done the diffractometer is brought to the expected $\vec{Q}$-vector and the sample φ angle must be scanned over a large range, depending on the symmetry of the ordering phenomena.

Prior knowledge of sample alignment is essential for experiments on thin films or single crystal samples. For single crystals pre-experiment lineup with a Laue diffractometer is necessary as scattering setups typically utilize point or limited 2D detectors. For films the orientation is typically dictated by the substrate, but pre-experiment confirmation is also essential. This is an area where beamtime is commonly underutilized in accessing the sample orientation. If aligned beforehand the variation is typically within a few degrees. However, if a clear surface cannot be achieved, then sample alignment before the experiment is much more critical and 2D detectors are nearly always required. Sample mosaicity or poor crystallinity (e.g. soft van der Waal crystals) lead to difficulty in finding a well-defined scattering orientation. One tool that can aid this is using a focused alignment laser to find sample regions with well-defined surface (provided sample cleaves well) that can be mapped under a microscope to determine a good starting location. Also, remember that diffractometers are not perfect so samples can move in the sub-100 um regime requiring realignment with large changes in diffraction condition. This is typically referred to as the sphere of confusion and varies greatly from instrument to instrument.

If following previously published work, care ought to be taken as crystals can be cut along a variety of directions and it might not be apparent in the publication what the sample orientation is. The same can be true for thin films which have a variety of substrate orientations. These details can be essential because in some cases, orienting a sample along a principal axis might not be possible since the peaks might not then be accessible in the diffractometer.

Sample thickness and related penetration depth of the X-rays at resonant energy, especially for thin film samples will have an impact on the experimentally measured sample volume. For single crystals of micron(s) thickness this is typically less important but can still be important for very high X-ray energies. The penetration depth is a function of the atomic species, with higher Z giving stronger absorption, the density of the material, and the X-ray energy. A great tool for calculating the estimated penetration depth for a particular sample is available in the X-ray database available at the Lawrence Berkeley Laboratory Center for X-ray Optics (cxro.lbl.gov) [20, 21]. However, this database does not contain detailed information about optical constants near resonances, which can lead to a much shorter penetration depth. If one has X-ray absorption spectroscopy data, it can be scaled to the non-resonant tables to get a more accurate number. Generally speaking, at soft X-ray energies one can expect penetration depths of 50–500 nm for normal incidence while in the hard X-ray regime ranges of 2–200 μm are common. Thus, with soft X-rays probing buried layers can be problematic depending on capping layer thickness. Furthermore, self-absorption can be a significant factor for the resonance shape.

For hard X-rays the higher penetration depth can severely limit the scattering signal from thinner samples. The incidence angle can be varied to remedy this to an extent, by increasing the apparent thickness of the sample as a factor of $\frac{1}{\sin\eta}$. Care must be taken as this also increases the beam footprint proportionally. Normally incident angles of ~ 5° can be used, giving approximately a 10-fold increase in apparent sample thickness. In extreme cases with large enough samples and focused beams incident angles of 1° can be used to get a ~ 50-fold larger apparent thickness. Below this angle you start to approach the critical angle, which depends on several factors but is typically between 0.1 and 1° for synchrotron X-rays on solids. The ability to reach such incidence angles for a given $\vec{Q}$-vector depends strongly on the experimental apparatus and is not always possible. A final important factor is the sample dimensions. Beam size is highly beamline dependent and can range from mm to nm depending on X-ray optics. This factor tends to be more important for single crystal samples with synthesis limited dimensions. While focused beams can be used for smaller samples, X-ray beam damage, especially in vacuum environments, can become important, especially in the soft X-ray regime, where the absorption lengths are shorter and can result in more energy deposited per sample volume.

2.2 Resonant Scattering Measurements

We begin with an important note that the peak labels, e.g. (001), are determined by the space group symmetry of the system under study. However, this is often not the case. Using a simpler pseudocubic unit cell rather than using the true orthorhombic or rhombohedral symmetries typically simplifies the experiment and is often done [22]. While important to determine which space group will be used for the experiment for convenience, it is critical eventually to attempt to resolve the actual symmetry. Calculations of sample orientation with respect to 4 or 6 circle diffraction measurements are more specialized than simple powder diffraction, given the large number of angles with multiple existing solutions [23, 24]. It will be necessary to make some decisions about which angles to fix and the diffraction software, e.g. packages such as SPEC (Certified Scientific Software, Inc.), will have multiple options for these cases. This is most important when considering how angles will rotate to the desired $\vec{Q}$ position, as it may be mechanically impossible or create a collision between circles.

After the sample is aligned and the diffractometer is brought to the approximate ordering phenomena $\vec{Q}$-vector, non-$\vec{Q}$ resolved florescence measurement across the resonant edge can be performed to confirm the energy calibration. This experiment requires a monochromated X-ray beam and there will always be variations in the exact energy due to the specifics. Therefore, it is best to check the position of the resonance in X-ray absorption, which can be accomplished with a separate fluorescence detector or sometimes positioning the area detector away from the Bragg condition. This also highlights that fluorescence backgrounds can be an issue when scattering intensity is weak. Resonance phenomena are typically strongest close to the white line of the fluorescence, which serves as a good guide to the starting conditions for the photon energy.

After finding the peak of interest, an energy scan at fixed $\vec{Q}$ should be done to confirm the reflection has a resonant response. Typical resonant response shows a strong enhancement of the scattering intensity and total loss of intensity away from the resonance. Ideally in a resonant scan, one seeks out peaks that maximize the contribution of the order parameter, as we will show in the examples below. However, in cases of orbital or magnetic ferro-type ordering the resonance phenomena can be present but difficult to resolve due to the much weaker contribution to the overall reflection that sits at the same $\vec{Q}$ as the lattice. In these cases, ferromagnetic ordering can be resolved by use of polarization analysis, but even at resonance the magnetic signal is mostly still too weak to extract.

As has been noted and you will see in the examples, polarization analysis is key in these experiments. For the hard X-ray regime, this is often a standard item at the beamlines. It is possible in the soft X-ray regime [25, 26], although not as commonly used due to weak signal and complexity of the instrument that is all in vacuum. An issue if the signal is weak is that the polarimeter is not perfect and a small leakage of signal from the non-polarization rotated channel can occur. Additionally, IDs generate significant higher order energy flux, if the peak of interest is a third order of a strong lattice reflection, leaked higher energy X-rays (higher odd multiple orders) and in particular with Si (111) oriented monochromators will generate significant 3rd order flux overwhelming weak magnetic or charge difference reflections from such a higher order structural reflection that a polarization analyzer can't sufficiently suppress. As discussed, most beamlines have optical mirrors specifically to eliminate higher order flux but care needs to be taken when looking for very weak commensurate reflections. These can be identified by the intrinsic peak width; higher energy generated peaks will be significantly sharper.

With confirmation of the resonance dependence, accurate scans through calculated reciprocal space can be achieved, sometimes termed *hkl* scans. The width of the peak along each vector direction corresponds to the correlation length of the phenomena in that direction. In the case of macroscopic crystals this can reveal important information about the dimensional behavior of the ordering. However, the resolution of the instrument can limit the sizes that are measurable. Typically, this will be on the order of microns. In the case of thin films, the growth direction correlation length is intrinsically limited by the thickness of the film creating finite size effects in reciprocal space extending to large volumes of volume, furthermore the near zero space of interfaces between substrate and air or different materials create boundary conditions that can further limit order parameter correlations orthogonal to interface. However, this phenomenon offers opportunities to study the intricate nature of interfaces themselves and the exotic physics that emerge from engineered heterostructured films.

Magnetic: Measurement of magnetic ordering with resonant X-ray scattering is a leading probe for non-ferromagnetic systems in particular, where the total magnetic moment often goes to zero, as in an antiferromagnetic system, or is quite small, such as canted antiferromagnetic or ferrimagnetic systems. Here the usual magnetization probes can become ineffective due to the much lower signal-to-noise ratio, particularly for the case of thin films. We focus here on the second term of Eq. 15, noting that in certain cases the higher order terms can be relevant [13]. The distinct advantage of resonant magnetic scattering is that the antiferromagnetic structure necessarily differs in periodicity in comparison to the structure lattice. In the simplest case of A-type antiferromagnetism where in-plane ferromagnetic layers flip direction for each layer in the out-of-plane direction, see Fig. 3, this leads to a doubling of the unit cell along the out-of-plane direction in real space, giving peaks in reciprocal space half-way between structure peaks, typically referred to as half-order peaks. More complex orderings also produce such reflections along the direction in reciprocal space corresponding to their ordering direction. The importance here is that despite the relatively low magnetic scattering cross-section (typically 10^{-4} compared to structural reflections) the absence of structural reflections in the vicinity makes these features readily observable given a strong enough X-ray source. Despite the feasibility of these measurements with modern 3rd generation synchrotron and X-ray free electron laser sources, some special conditions are still required to be met when performing such experiments.

Strong consideration must be given to the orientation of the sample's magnetic moment when setting up magnetic scattering experiments. As discussed above, the magnetic scattering cross-section is proportional to the dot product of the direction of the magnetic moments with the cross product of the incoming and outgoing photon polarization. The magnetic scattering in the cross-polarization channel changes the linear polarization by $\frac{\pi}{2}$ giving a cross product that lies in the direction of the scattered beam, for vertical scattering geometry (see Fig. 2), or the incident beam, for horizontal scattering geometry. Here horizontal (vertical) refers to the typical synchrotron radiation condition of X-ray polarization parallel (normal) to the ground. For the cross-product to lie along the incident or scattered beam the scattering plane must then be

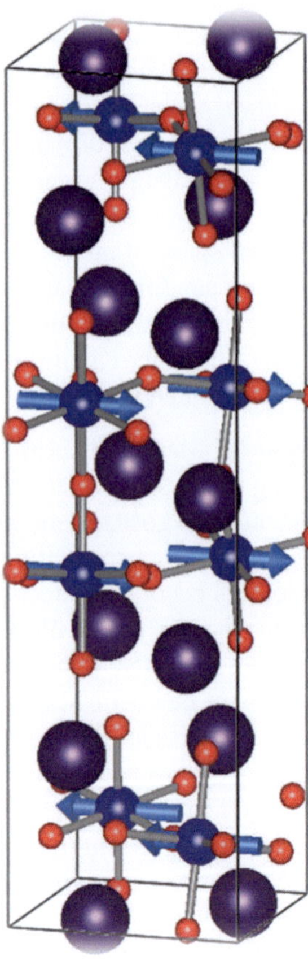

Fig. 3 Magnetic structure of $Ca_3Ru_2O_7$ as an example of simple A-type AFM order [27]

entirely horizontal or vertical, otherwise the cross product can lie in any direction. Due to the dependence of the cross-section on the dot product, care must be taken when setting up the experiment to ensure the sample orientation does not align the moments 90° from the scattered (vertical scattering plane) or incident (horizontal scattering plane) beam. Due to this factor, the intensity of the different magnetic reflections can vary quite strongly based on all the alignment angles. The necessity of the resonance condition X-ray energy typically limits which peaks are accessible, especially in the case of soft X-rays, making calculation of the cross-sections for each reflection very helpful to ensure a successful experiment.

Despite these considerations, in the case of very small magnetic moments, very thin films (crystal truncation rods), weaker X-ray sources, insufficient shielding, or other complications, the background signal can overwhelm the magnetic scattering. In this case polarization dependence can be used to filter out the un-flipped horizontally-polarized light. Note as discussed in Sect. 1, the polarization behavior is typically assumed to match the cubic case, but one must be careful when measuring lower symmetry systems, where the cross-polarization channel is not necessarily entirely magnetic in origin, see the next section on orbital scattering.

Orbital: Orbital ordering (OO) can also arise in quantum materials and is typically harder to probe. However, near a resonance the anisotropy in the electronic structure can result in a clear contrast even in the case where the lattice sites are locally structurally equivalent (e.g. identical B–O distances). Manganites are probably the most explored in this case [28], but some more recent examples have appeared such as the layered ruthenates (e.g. Ca_2RuO_4), as is shown in Fig. 4. Here the orbital ordering is from an alternating arrangement in- versus out-of-plane t_{2g} orbitals arising from strong interactions [29, 30]. In this case, the orbital order can be probed with resonant scattering due to the anisotropy in the electronic structure which doubles the unit cell, as will be discussed in the example below.

In this case, the anisotropy in the local electronic environment can lead to scattering conditions for specific peaks that cause a polarization rotation upon scattering, like the magnetic case. In this case the assumptions of Eq. 11 are not valid and the scattering length cannot break down into the two simple terms of Eq. 15. Thus, the cross-polarization channel can have contributions from non-magnetic scattering. Using the same geometry with a polarization analyzer, these terms can be measured.

Charge: Ordering of the charge, also called charge disproportionation or a charge-density wave, is a key-phenomena observed in superconducting cuprates and several other materials [31–35]. Here, structurally equivalent atoms obtain different valence states, forming a new larger real space lattice in analogy to the antiferromagnetic ordering discussed above (Fig. 5). Importantly, such ordering can be quite long wavelength, as in the cuprates, due to organization of the missing (or added) charge into stripe features. Such features are also often incommensurate with the underlying lattice, i.e. they do not repeat as integer multiples of the underlying crystal lattice. In other systems commensurate ordering is commonly observed leading to half order peak in analogy with anti-ferromagnetic ordering [36]. This means the $\vec{Q}$-vector for charge ordering can appear throughout reciprocal space.

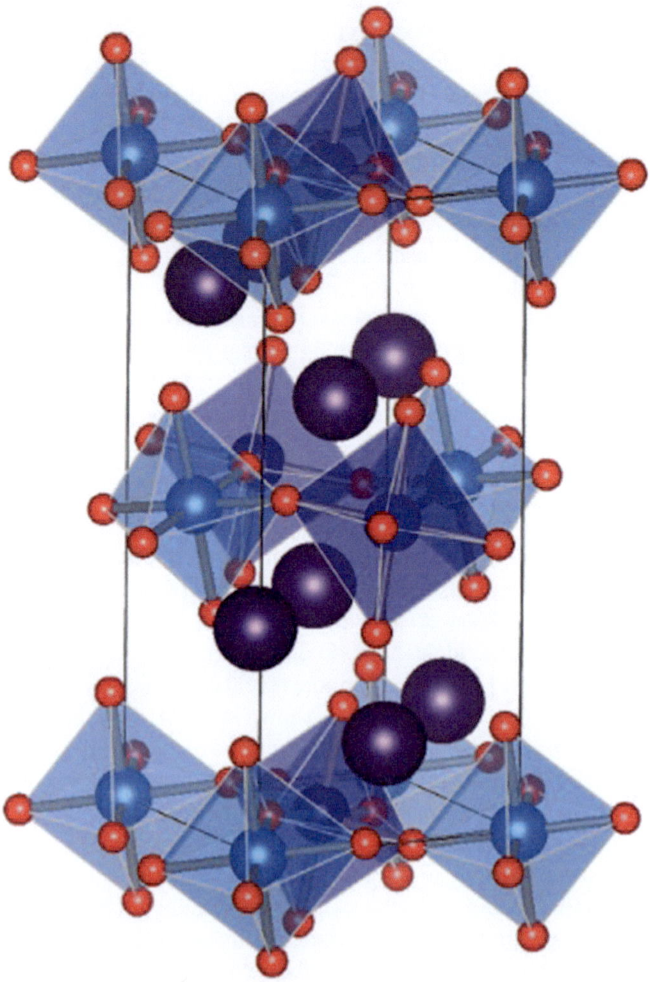

Fig. 4 Orbital ordering in Ca_2RuO_4 showing sites with different t_{2g} orbital orientations (light and dark blue) that doubles the unit cell along the a-axis direction when probed at resonance [29]

Charge order (CO) scattering originates from the first term of Eq. 15, assuming a reasonably cubic crystal environment. Thus, cross polarization analysis can be used to distinguish charge order scattering from magnetic scattering. As usual, caution must be used in systems where this assumption is far from realistic.

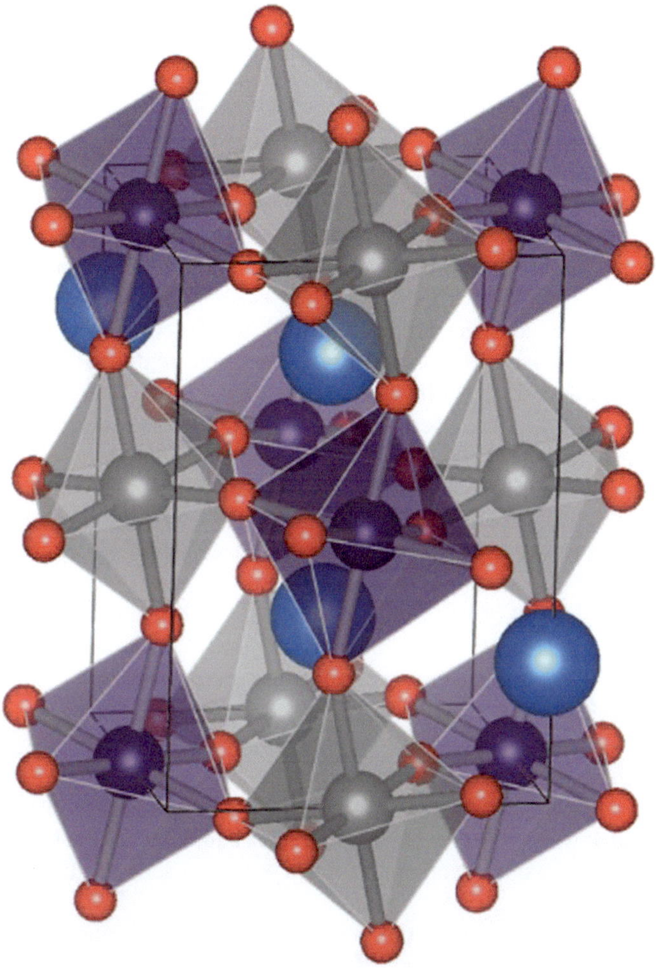

Fig. 5 Charged ordered phase of $NdNiO_3$ consisting of alternating short (grey) and long bond sites (purple) associated with two distinct local charge distributions [35, 36]

3 Case Studies

3.1 Resonant Magnetic Scattering

As discussed above resonant scattering is an extremely powerful technique for the measurement of antiferromagnetic ordering due to the change in symmetry creating reflections with purely magnetic origin. This technique is useful for crystals which cannot be made large enough for magnetic neutron scattering measurements or heavily neutron absorbing rare-earth materials. A famous example of demonstrating the power and importance of this technique is in the case of the Ruddlesden–Popper iridate materials, in particular Sr_2IrO_4.

High temperature superconductivity in the copper-bearing oxides (cuprates) has been a major focus of condensed matter research over the past four decades. Due to the complexity of the interactions in the material, the nature of the cooper pairing mechanism has remained elusive. Therefore, much effort has been made to find and study analogs to the cuprates which share structural and ordering similarities to help shed light on the necessary ingredients for this intriguing phenomenon. Very shortly after the initial discovery of superconductivity in the cuprates, Sr_2RuO_4 was also found to superconduct, albeit at much lower temperatures. It was believed the structural, electronic, and magnetic similarities to superconducting $La_{2-\delta}Sr_\delta CuO_4$ could explain such behavior. In particular, the electronic configuration of the Ru ion, with one hole in the outer t_{2g} orbital, was similar to the one hole in the e_g band of the Cu ion. This led to much speculation about other possible superconductors that could be realized using similar lattice and electronic structure.

One such compound that was pursued as a potential superconductor was Sr_2IrO_4 [37]. By the combination of crystal field splitting and the Russell–Saunders coupling scheme this was also expected to have a single hole in the t_{2g} xy orbital, analogous to Sr_2RuO_4 [27]. However, it was quickly found that this compound does not superconduct despite all of the correct "ingredients" being in place.

Resolution of this puzzle came about through the use of resonant magnetic scattering at the (1 0 22) reflection of a small single crystal of Sr_2IrO_4 [37]. Resonant energy scans at both the L_3 and L_2 edges for this reflection showed almost no signal for the L_2, while the L_3 showed a strong response (Fig. 6). In the orbital schemes used for the cuprates and Sr_2RuO_4 this was impossible, as the transition rules give an equal number of available electrons for both the $2p_{3/2}$ and $2p_{1/2}$ transitions into the xy orbital, which will lead to the observation of an equal intensity at both edges. The authors put forward a new theory, now known as the J_{eff} scheme, that could explain this anomalous behavior.

Strong spin-orbit coupling in the Ir ions, absent in the Cu ions and, to a lesser extent, in the Ru ions, leads to a new electronic configuration. Here the energy of the spin-orbit coupling is on the order of the crystal field splitting, and so both effects must be incorporated to get a true picture of the orbital configuration. Instead of the pure t_{2g} and e_g bands of the other compounds, this leads to an upper $J_{eff} = \frac{1}{2}$ and lower $J_{eff} = \frac{3}{2}$ band. The $J_{eff} = \frac{1}{2}$ band is then half filled, leaving a single hole analogous to the cuprate and Sr_2RuO_4 situation [37]. However, the $J_{eff} = \frac{1}{2}$ orbital configuration is completely different from the other compounds, which along with the strong influence of spin-orbit coupling, is likely to blame for the lack of superconductivity in this compound.

Beyond this exciting new model, the authors were also able to measure the intensity of multiple magnetic reflections with and without large applied magnetic field to solve another mystery of SrI_2O_4. Under large magnetic field it was found that a small ferromagnetic ordering could be obtained. However, neutron scattering was not able to detect the magnetic ordering, so this behavior could not be explained. By modeling

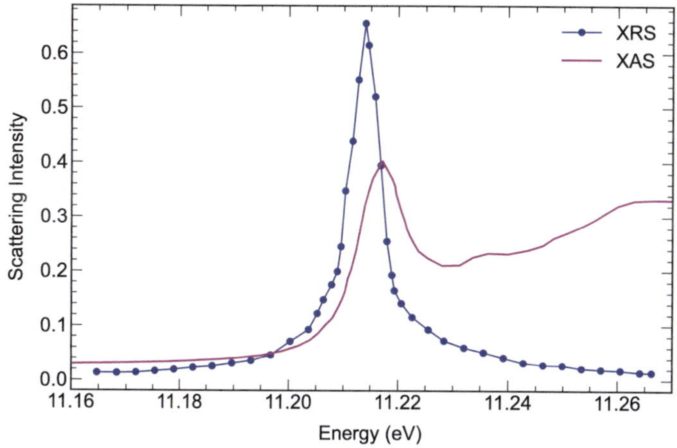

Fig. 6 Resonant absorption (magenta) and magnetic scattering (blue) at the Ir L_3 edge on Sr_2IrO_4. Data adapted from Ref. [37]

the ordering from the observed magnetic reflections using magnetic scattering, the authors were able to explain this behavior as a transition as originating from the alignment of the canted AFM moments in each IrO_2 layer, which are anti-aligned in the absence of a magnetic field.

3.2 *Resonant Orbital Scattering*

The ordering of the orbital degree of freedom is another fundamental aspect of quantum materials. Orbital order (OO) is typically the hardest to probe and is closely related to magnetic and charge order in many cases. Here we use the example of Ca_2RuO_4 to explore how to probe the orbital state [29, 30]. As shown in Fig. 4, the orbitally ordered state doubles the unit cell along the *a*-axis. This gives rise to a structurally forbidden peak (100), which unfortunately is at the same wave vector as the AFM order peak (see Fig. 7). Fortunately, in this case, the AFM ordering occurs at a lower temperature than the OO so it is possible to separate the contributions from both AFM and OO (see Fig. 7). It is apparent that the OO contribution has a different resonant shape due to the different origin.

To elaborate on this further, let's consider the different contributions. For the AFM order, as discussed above, the intensity arises from the f^1 term. The orbital terms are from the charge anisotropy along the different crystalline axes that are in the f^0 terms along the diagonal of the scattering factor matrix. To measure these terms, it is important to consider the scattering geometry. If the scattering plane is aligned with the principal axis showing a charge anisotropy, there will be no contribution to the $\sigma'\pi$ channel. In order to measure the OO terms, it is necessary to have some way to break this condition. This can occur either due to the octahedral tilting out of the scattering plane [27] or from an off-specular scattering condition [38].

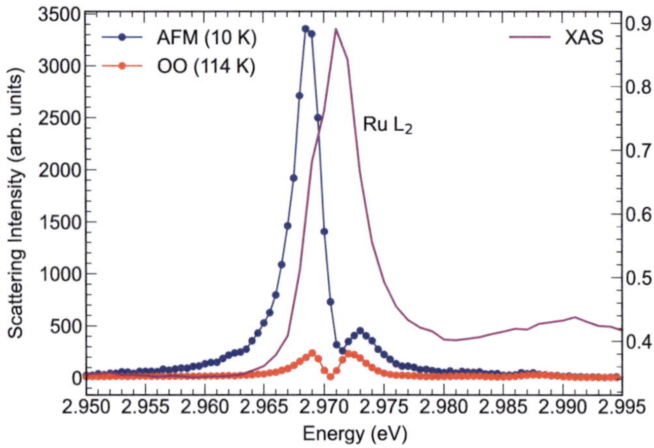

Fig. 7 (1 0 0) resonant scattering above and below the AFM ordering temperature showing the sensitivity to t_{2g} orbital ordering at the same wave vector. The right axis shows the XAS data for comparison. Data adapted from Ref. [29]

3.3 Resonant Charge Scattering

Resonant scattering can also be incredibly useful for observing subtle differences in the charge occupation of ions due to an ordering of charge into $\pm\ \delta$ sites. This is due to the direct interaction of the X-rays and charge cloud surrounding each ion. In the case of the perovskite nickelates, early neutron scattering showed evidence for a monoclinic distortion and charge ordering for the most distorted nickelate, $YNiO_3$, but not the other compounds featuring larger rare earths in place of Y (Nd, Pr, Eu, etc.). However, these compounds (with the exception of $LaNiO_3$) all shared similar metal-insulator transitions (MITs), leading to speculation they may all share the same charge-ordered ground state as in $YNiO_3$.

Staub et al. proved this ground state is indeed ubiquitous to insulating perovskite nickelates using resonant X-ray scattering in the case of bulk $NdNiO_3$ [36]. New reflections corresponding to a rock-salt type ordering were observed at temperatures below the MIT. Energy scans at these reflections showed strong resonance at the Ni K-edge, showing conclusively the ordering originated from differences in charge between neighboring Ni sites. The authors show that the structure factor for this reflection depends directly on the difference of the Ni site form factor. Thus, when they are equivalent no reflection is observed, but when they become charge ordered a reflection arises. Furthermore, this dichroic behavior leads to a strong azimuthal dependence, with the (015) having an inverse resonant profile. The authors were able to observe this directly.

This was subsequently applied to many heterostructures of nickelates, where one was trying to use dimensionality to manipulate the ordered phase [33, 39]. Following the example set forth by Staub et al., Fig. 8 shows the data from a $EuNiO_3/LaNiO_3$ superlattice orthorhombic (011) reflection. There is a very strong enhancement below the insulator to metal transition where the charge ordered state appears (see Fig. 5). The temperature dependence is shown in the bottom panel. Note that the signal does not vanish above the transition to the metallic state. Recent experiment and theory showed that lattice distortions and tilts can also lead to a resonant signal depending on the peak measured [40].

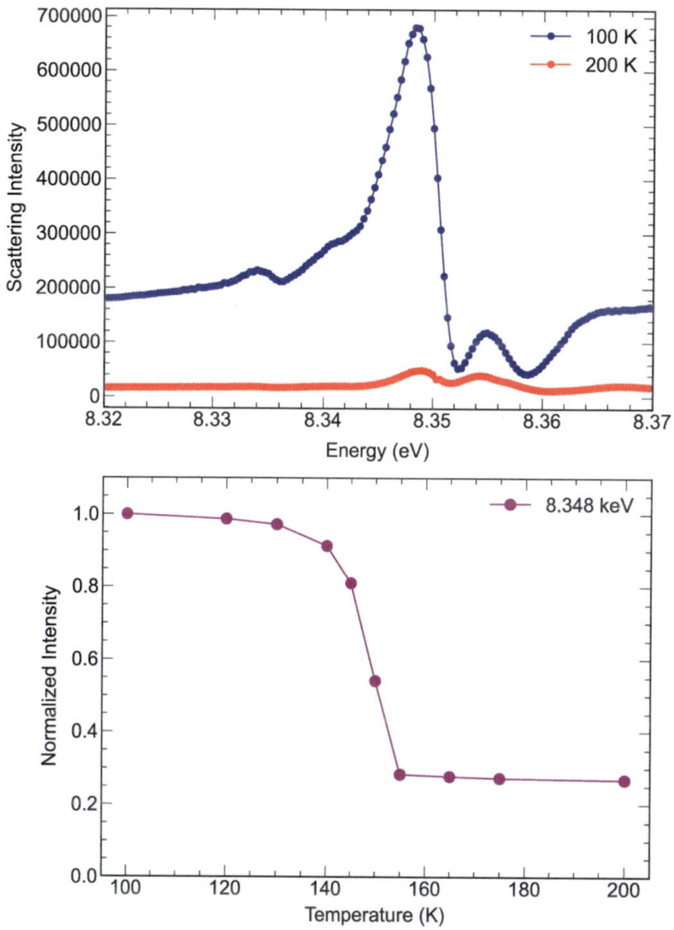

Fig. 8 Top panel: Resonant scattering signal from the $(0\ 1\ 1)_{or}$ reflection of the nickelate superlattice showing the charge resonance behavior at the Ni K-edge. Bottom panel: Order parameter measurement showing the collapse of charge order at the insulator to metal transition. Data adapted from Ref. [39]

4 Summary—Outlook

Here we discuss how such resonant scattering techniques will evolve with the coming improvements of X-ray light sources. On the synchrotron front, light sources around the world are harnessing new accelerator technologies to creates diffraction limited storage rings that shrink the emittance of the electron beam and enhance the brilliance of by orders of magnitude [41] enhancing the achievable coherence that will open new opportunities for resonant scattering. Currently, nanoscale resolution coherent imaging using resonant scattering has utilized strong scattering at the soft energy transition metal L edges [42, 43]. However, due to the long wavelengths this requires special cases with large electronic and magnetic unit cells accessible in angular space. With the enhanced brilliance of these coming upgrades, new capabilities will allow these same experiments in the hard X-ray regime. Recently some nanoscale resonant diffraction has been done [44], but most other imaging in resonant diffraction mode in the hard X-ray regime has been limited to micron scale or larger [45–47]. Pushing nanoscale probes in the hard X-ray regime will enable more insight to detailed nanoscale domain imaging.

Another frontier is tapping into the X-ray coherence to use techniques such as X-ray Photon Correlation Spectroscopy (XPCS) [48] or phase-retrieval methodologies such as Bragg Coherent Diffraction Imaging (BCDI) [49]. XPCS is utilized to follow fluctuations in materials by tracking the scattering of coherent X-rays. When the sample has spatial and temporal modulations of order, this can imprint on the scattering of X-rays and lead to a speckle pattern. By tracking a series of snapshots, these are used to construct a correlation function in terms of Q and time that contains insight into the spatial and temporal nature of the scattering [48]. However, as in the case of coherent imaging at resonant conditions, almost all of the studies are in the soft X-ray regime, but this will change with the next wave of light source upgrades. The rapidly developing use of resonant effects in coherence studies is also described in Chapter 3 of this book.

Up to this point, we have focused on synchrotron-based sources. X-ray free electron lasers (XFELs) are also opening new doors for resonant scattering. XFELs operate by taking an electron bunch from a linear accelerator and feeding into a long chain of undulators. This results in a process of amplification akin to an optical laser and creates X-rays pulse of very high intensity with temporal widths in the < 100 fs domain. Contrast this with pulses from a synchrotron that have ~ 100 ps widths. To date this has enabled a few studies of electronic and magnetic order by working at hard X-ray resonances in both elastic [50] and inelastic modalities [51]. With the current movement of XFELs from warm to superconducting linear accelerators, this will enhance signal to noise by increasing from ~ 100 Hz to MHz X-ray pulse repetition rates.

Questions and Problems (Resonant Hard X-Ray Scattering Probes of Electronic and Magnetic Order: A Practical Guide)

Q 5.1. Plot the spin azimuthal angle orientation dependence of the magnetic scattering amplitude in the $\pi'\pi$ and $\pi'\sigma$ channels of S_M.

i. Which of these terms is important in charge-magnetic interference scattering?
ii. Plot the expected spin orientation dependence of the intensity of charge-magnetic interference scattering and indicate which spin orientation produces the highest intensity.

Q 5.2. The photon-energy dependence of the magnetic scattering amplitudes or intensities for Ir and Ru are shown in Figs. 1 and 7, respectively.

i. Examine these plots and estimate how monochromatic the X-ray beam must be to produce reliable magnetic scattering information. Express your answer in eV.
ii. What is the fractional energy resolution required in the X-ray monochromator, i.e. the energy width divided by the total energy?
iii. In the hard X-ray regime monochromators based on Si (111) reflections offer an energy resolution on the order of slightly worse than 10^{-4}. Is a Si (111) monochromator sufficient for the experiments shown in Figs. 1 and 7?

Q 5.3. Diffraction from magnetic order requires that the X-ray wavelength be sufficiently short to allow the X-ray beams to meet the Bragg condition. The antiferromagnetic order in rare-earth nickelates arises at ¼ ¼ ¼, see for example Fig. 5 and the study of $NdNiO_3$ in Song et al. [52] work. $NdNiO_3$ has cubic symmetry and lattice parameter (for the purposes of this rough estimate) of approximately 4 Å. Can a resonant scattering experiment at a photon energy of 853 eV, matching the Ni L_3 edge, reach this reflection?

Q 5.4. Consider the polarization analysis scheme in the scattering diagram in Fig. 2.

i. Sketch the orientation of the polarization analyzer crystal in Fig. 2 that would be required to allow the $I_{\sigma\pi}$ channel to be measured.
ii. What Bragg angle is required to achieve the maximum suppression of the $I_{\sigma\sigma}$ channel?
iii. Explain why it might be challenging to find an analyzer crystal that satisfies this condition for an arbitrary photon energy, i.e. one matching a particular resonant edge.

References

1. Y. Tokura, N. Nagaosa, Orbital physics in transition-metal oxides. Science **288**, 462 (2000)
2. B. Keimer, J.E. Moore, The physics of quantum materials. Nat. Phys. **13**, 1045 (2017)
3. G. Subías, J. García, J. Blasco, J. Herrero-Martín, M.C. Sánchez, Resonant X-ray scattering in 3d-transition-metal oxides: anisotropy and charge orderings. J. Phys. Conf. Ser. **190**, 012085 (2009)
4. T.A.W. Beale et al., REXS contribution to electronic ordering investigation in solids. Eur. Phys. J. Spec. Top. **208**, 89 (2012)
5. C. Vettier, Resonant elastic X-ray scattering: where from? Where to? Eur. Phys. J. Spec. Top. **208**, 3 (2012)
6. J. Fink, E. Schierle, E. Weschke, J. Geck, Resonant elastic soft X-ray scattering. Rep. Prog. Phys. **76**, 056502 (2013)
7. L.J.P. Ament, M. Veenendaal, T.P. Devereaux, J.P. Hill, J. Brink, Resonant inelastic X-ray scattering studies of elementary excitations. Rev. Mod. Phys. **83**, 705 (2011)
8. M. Blume, Magnetic scattering of X rays. J. Appl. Phys. **57**, 3615 (1985)
9. J.P. Hill, D.F. McMorrow, X-ray resonant exchange scattering: polarization dependence and correlation function. Acta Crystallogr. Sect. A Found. Crystallogr. **52**, 236 (1996)
10. E. Beaurepaire, H. Bulou, F. Scheurer, J.-P. Kappler (eds.), *Magnetism: A Synchrotron Radiation Approach*, vol. 697 (Springer Berlin Heidelberg, Berlin, Heidelberg, 2006)
11. J.J. Sakurai, *Advanced Quantum Mechanics* (Addison-Wesley, Reading, MA, 1980)
12. R.L. Kronig, On the theory of dispersion of X-rays. J. Opt. Soc. Am. **12**, 547 (1926)
13. M.W. Haverkort, N. Hollmann, I.P. Krug, A. Tanaka, Symmetry analysis of magneto-optical effects: the case of X-ray diffraction and X-ray absorption at the transition metal L23 edge. Phys. Rev. B **82**, 094403 (2010)
14. S. Nandi et al., Strong coupling of Sm and Fe magnetism in SmFeAsO as revealed by magnetic X-ray scattering. Phys. Rev. B **84**, 054419 (2011)
15. J.O. Cross, M. Newville, J.J. Rehr, L.B. Sorensen, C.E. Bouldin, G. Watson, T. Gouder, G.H. Lander, M.I. Bell, Inclusion of local structure effects in theoretical X-ray resonant scattering amplitudes using ab initio X-ray-absorption spectra calculations. Phys. Rev. B **58**, 11215 (1998)
16. D. Haskel, G. Fabbris, M. Zhernenkov, P.P. Kong, C.Q. Jin, G. Cao, M. Veenendaal, Pressure tuning of the spin-orbit coupled ground state in Sr_2IrO_4. Phys. Rev. Lett. **109**, 027204 (2012)
17. J. Strempfer, S. Francoual, D. Reuther, D.K. Shukla, A. Skaugen, H. Schulte-Schrepping, T. Kracht, H. Franz, Resonant scattering and diffraction beamline P09 at PETRA III. J. Synchrotron Radiat. **20**, 541 (2013)
18. K. Momma, *VESTA: Visualization for Electronic and STructural Analysis* (n.d.)
19. D. Palmer, S. Palmer, *CrystalMaker Software* (n.d.)
20. B.L. Henke, E.M. Gullikson, J.C. Davis, X-ray interactions: photoabsorption, scattering, transmission, and reflection at E = 50–30,000 eV, Z = 1–92. At. Data Nucl. Data Tables **54**, 181 (1993)
21. B.L. Henke, E.M. Gullikson, J.C. Davis, *X-Ray Interactions with Matter*, https://henke.lbl.gov/optical_constants/
22. M. Brahlek, A.K. Choquette, C.R. Smith, R. Engel-Herbert, S.J. May, Structural refinement of Pbnm-type perovskite films from analysis of half-order diffraction peaks. J. Appl. Phys. **121**, 045303 (2017)
23. W.A. Wooster, On the use of a four-circle X-ray diffractometer. J. Sci. Instrum. **42**, 685 (1965)
24. M. Lohmeier, E. Vlieg, Angle calculations for a six-circle surface X-ray diffractometer. J. Appl. Crystallogr. **26**, 706 (1993)
25. U. Staub, V. Scagnoli, Y. Bodenthin, M. García-Fernández, R. Wetter, A.M. Mulders, H. Grimmer, M. Horisberger, Polarization analysis in soft X-ray diffraction to study magnetic and orbital ordering. J. Synchrotron Radiat. **15**, 469 (2008)
26. T.A.W. Beale, T.P.A. Hase, T. Iida, K. Endo, P. Steadman, A.R. Marshall, S.S. Dhesi, G. Laan, P.D. Hatton, RASOR: an advanced instrument for soft X-ray reflectivity and diffraction. Rev. Sci. Instrum. **81**, 073904 (2010)

27. B. Bohnenbuck et al., Magnetic structure and orbital state of $Ca_3Ru_2O_7$ investigated by resonant X-ray diffraction. Phys. Rev. B **77**, 224412 (2008)
28. J. Herrero-Martín, J. García, J. Blasco, G. Subías, Mn K edge resonant X-ray scattering of half-doped manganites. Eur. Phys. J. Spec. Top. **208**, 107 (2012)
29. I. Zegkinoglou et al., Orbital ordering transition in Ca_2RuO_4 observed with resonant X-ray diffraction. Phys. Rev. Lett. **95**, 136401 (2005)
30. S. Mohapatra, A. Singh, Magnetic reorientation transition in a three orbital model for Ca_2RuO_4—interplay of spin-orbit coupling, tetragonal distortion, and Coulomb interactions. J. Phys. Condens. Matter **32**, 485805 (2020)
31. R. Comin, A. Damascelli, Resonant X-ray scattering studies of charge order in cuprates. Annu. Rev. Condens. Matter Phys. **7**, 369 (2016)
32. J.M. Tranquada, Charge stripes and antiferromagnetism in insulating nickelates and superconducting cuprates. J. Phys. Chem. Solids **59**, 2150 (1998)
33. S. Middey, J. Chakhalian, P. Mahadevan, J.W. Freeland, A.J. Millis, D.D. Sarma, Physics of ultrathin films and heterostructures of rare-earth nickelates. Annu. Rev. Mater. Res. **46**, 305 (2016)
34. M.B. Salamon, M. Jaime, The physics of manganites: structure and transport. Rev. Mod. Phys. **73**, 583 (2001)
35. J.L. García-Muñoz, M.A.G. Aranda, J.A. Alonso, M.J. Martínez-Lope, Structure and charge order in the antiferromagnetic band-insulating phase of $NdNiO_3$. Phys. Rev. B **79**, 134432 (2009)
36. U. Staub, G.I. Meijer, F. Fauth, R. Allenspach, J.G. Bednorz, J. Karpinski, S.M. Kazakov, L. Paolasini, F. d'Acapito, Direct observation of charge order in an epitaxial $NdNiO_3$ film. Phys. Rev. Lett. **88**, 126402 (2002)
37. B.J. Kim, H. Ohsumi, T. Komesu, S. Sakai, T. Morita, H. Takagi, T. Arima, Phase-sensitive observation of a spin-orbital Mott state in Sr_2IrO_4. Science **323**, 1329 (2009)
38. T. Kiyama, Y. Wakabayashi, H. Nakao, H. Ohsumi, Y. Murakami, M. Izumi, M. Kawasaki, Y. Tokura, Resonant X-ray scattering in perovskite manganite superlattice—observation of "orbital superlattice." J. Phys. Soc. Jpn. **72**, 785 (2003)
39. S. Middey, D. Meyers, M. Kareev, Y. Cao, X. Liu, P. Shafer, J.W. Freeland, J.W. Kim, P.J. Ryan, J. Chakhalian, Disentangled cooperative orderings in artificial rare-earth nickelates. Phys. Rev. Lett. **120**, 156801 (2018)
40. Y. Lu et al., Quantitative determination of bond order and lattice distortions in nickel oxide heterostructures by resonant X-ray scattering. Phys. Rev. B **93**, 165121 (2016)
41. C.G. Schroer et al., PETRA IV: the ultralow-emittance source project at DESY. J. Synchrotron Radiat. **25**, 1277 (2018)
42. M. García-Fernández, S.B. Wilkins, M. Lu, Q. Li, K.E. Gray, H. Zheng, J.F. Mitchell, D. Khomskii, Antiferromagnetic domain structure in bilayer manganite. Phys. Rev. B **88**, 075134 (2013)
43. J. Li et al., Scale-invariant magnetic textures in the strongly correlated oxide $NdNiO_3$. Nat. Commun. **10**, 4568 (2019)
44. P.G. Evans et al., Resonant nanodiffraction X-ray imaging reveals role of magnetic domains in complex oxide spin caloritronics. Sci. Adv. **6**, eaba9351 (2020)
45. S. Boseggia, R. Springell, H.C. Walker, A.T. Boothroyd, D. Prabhakaran, D. Wermeille, L. Bouchenoire, S.P. Collins, D.F. McMorrow, Antiferromagnetic order and domains in $Sr_3Ir_2O_7$ probed by X-ray resonant scattering. Phys. Rev. B **85**, 184432 (2012)
46. S. Tardif, S. Takeshita, H. Ohsumi, J. Yamaura, D. Okuyama, Z. Hiroi, M. Takata, T. Arima, All-in–all-out magnetic domains: X-ray diffraction imaging and magnetic field control. Phys. Rev. Lett. **114**, 147205 (2015)
47. H. Ueda, Y. Tanaka, Y. Wakabayashi, T. Kimura, Observation of collinear antiferromagnetic domains making use of the circular dichroic charge-magnetic interference effect of resonant X-ray diffraction. Phys. Rev. B **98**, 134415 (2018)
48. O.G. Shpyrko, X-ray photon correlation spectroscopy. J. Synchrotron Radiat. **21**, 1057 (2014)

49. I. Robinson, R. Harder, Coherent X-ray diffraction imaging of strain at the nanoscale. Nat. Mater. **8**, 291 (2009)
50. P. Beaud et al., A time-dependent order parameter for ultrafast photoinduced phase transitions. Nat. Mater. **13**, 923 (2014)
51. M.P.M. Dean et al., Ultrafast energy-and momentum-resolved dynamics of magnetic correlations in the photo-doped Mott insulator Sr_2IrO_4. Nat. Mater. **15**, 601 (2016)
52. Q. Song et al., Antiferromagnetic metal phase in an electron-doped rare-earth nickelate. Nat. Phys. **19**, 522 (2023)

Measuring Octahedral Rotations in Perovskite Oxide Heterostructures

Steven J. May

Abstract This chapter describes how synchrotron diffraction can be used to measure octahedral rotations within ABO_3 perovskite oxide heterostructures. The collective rotations of the BO_6 octahedra displace the oxygen atoms from the face-centered positions, effectively doubling the perovskite unit cell. As a result, a series of half-order reciprocal lattice points arise, which can be measured from oxide films using synchrotron X-rays. The presence and absence of specific rotations is indicative of the rotation pattern and thus provides a direct means to determine the crystal structure of the thin film. Through quantitative analysis of the half-order peak intensities, the average position of the oxygen atoms can be refined in strained heterostructures, including ultrathin films and superlattices. The chapter provides step-by-step details regarding how to carry out this approach, including discussion of what peaks to measure, how to calculate peak intensities, and how to extract atomic positions through error-minimizing fits of measured and calculated intensities. Numerous examples are provided to demonstrate the described approach as well as to illustrate how these measurements can be used to provide insights into epitaxial arrangements, physical properties, and dynamic responses of perovskite films to electric fields and optical excitation.

Keywords Octahedral rotations · Crystal structure · Structural refinement · Perovskite heterostructures

1 Introduction

The tremendous diversity of properties exhibited by perovskite oxides is enabled by the wide range of chemistries that can be accommodated within the nominal ABO_3 formula and by the symmetry-lowering structural distortions that render the basic crystal structure flexible enough to accommodate an array of different *A*- and *B*-site compositions. The *A*-site is commonly occupied by alkali, alkali earth, and rare earth

S. J. May (✉)
Department of Materials Science and Engineering, Drexel University, Philadelphia, PA, USA
e-mail: smay@drexel.edu

D. Sando et al. (eds.), *X-Ray Scattering Techniques for Epitaxial Oxide Thin Films*,
https://doi.org/10.1007/978-981-96-5945-6_6

elements, while the B-site typically hosts $3d$, $4d$, or $5d$ transition metals as well as some group 13 and 14 elements. The feasibility of a given A-site cation and B-site cation forming the perovskite structure can be readily estimated through calculation of the tolerance factor, $t = \frac{(r_A+r_O)}{\left[\sqrt{2}(r_B+r_O)\right]}$ where r is the ionic radii of the A or B-site cation and the oxygen anion. The perovskite structure is generally stable between t values of approximately 0.8–1.0.

Behind this simple calculation lies a much more complex and interesting crystallographic situation. Perovskite does not refer to a single crystal structure but instead to a family of structures with a common inheritance of a corner-connected network of BO_6 octahedra but differing in the pattern and magnitude of octahedral rotations and distortions present within this network. Symmetry lowering A-site and B-site cation displacements also occur within some perovskite structures. As illustrated in Fig. 1a, b, BO_6 octahedral rotations result from displacements of the oxygen atoms from the ideal face-centered positions, resulting in a reduction of the B–O–B bond angles (θ) from 180° and an increase in the B–O bond lengths (d). The propensity for these rotations increases as t is reduced from its ideal value of 1. The physical meaning of $t < 1$ is that the A-site cationic radius is too small compared to the B-site cationic radius, causing the basic unit cell size to be smaller than that preferred by the B–O bond. Within this decreased lattice volume, the B–O bond length can avoid an energetically unfavored contraction through the emergence of collective rotations and tilts of the BO_6 octahedral units. These rotations arise in two types: in-phase rotations, in which the whole column of octahedral units rotate with the same handedness along a given axis, or out-of-phase, in which the octahedral units rotate with alternating handedness along an octahedral column.

The notation for categorizing the many rotation patterns was put forward by Glazer and is thus referred to as Glazer notation [1]. The rotations are considered about the three Cartesian directions with respect to the cubic (unrotated) perovskite structure. In-phase rotations are designated with a superscript "+", while out-of-phase rotations are indicated by a superscript "–". Figure 1c, d illustrates the difference between these in-phase and out-of-phase rotations. Rotations along each axis (a, b, or c) are then denoted by either $+$ or $-$. The absence of rotations is marked with a superscript "0". For instance, $a^-b^+c^0$ denotes a perovskite structure with out-of-phase rotations about the pseudocubic a-axis, in-phase rotations about the pseudocubic b-axis, and no rotations about the pseudocubic c-axis. If the rotational magnitudes are equal along different directions, then these axes are given the same letter in Glazer notation: $a^-a^-a^-$ indicates out-of-phase rotations that are equal in magnitude along the three axes. Numerous structural variants with differing symmetries result from the combinatorial possibilities of the different rotation types along different rotation axes [2–4].

The rotational patterns and magnitudes are not just interesting from a crystallographic viewpoint but are essential to understanding and designing physical properties within perovskites. As the rotations increase in magnitude (as θ is decreased), electronic bandwidth of the valence and conduction bands decreases resulting in changes to macroscopic physical properties such as metal–insulator transitions in

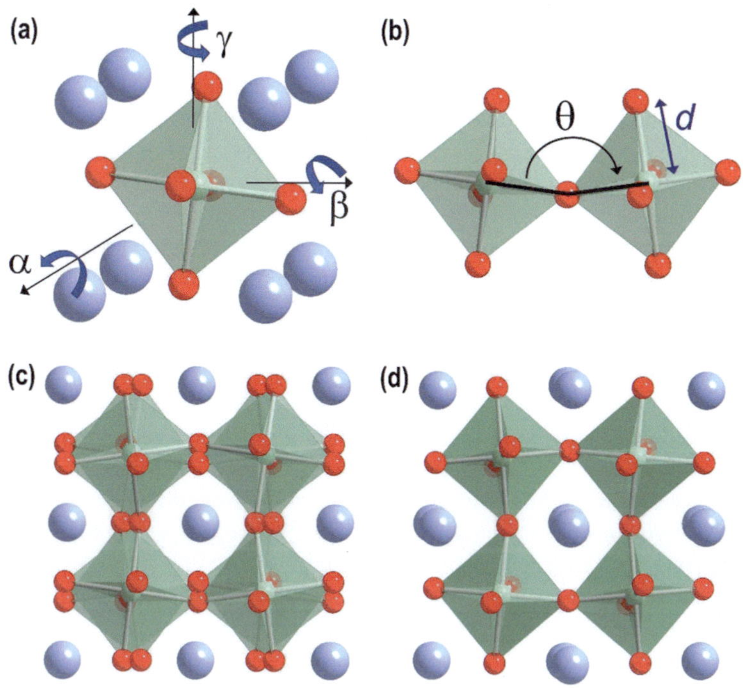

Fig. 1 Schematics of BO_6 octahedral rotations in ABO_3 perovskites. **a** The α, β, and γ rotational axes lay along the Cartesian directions of the pseudocubic perovskite unit cell displacing oxygen atoms (red spheres) from ideal face-centered positions. **b** The rotations modify the geometry of the BO_6 octahedral network by decreasing the B–O–B bond angle denoted by θ and increasing the B–O bond lengths denoted by d. **c** An illustration of the $a^-a^-c^+$ rotation pattern as viewed down an out-of-phase (a^-) axis. **d** An illustration of the $a^-a^-c^+$ rotation pattern as viewed down the in-phase axis (c^+)

nickelates and ferrites [5, 6], ferro-to-antiferromagnetic transitions in manganites [7], changes to magnetic ordering temperatures [8, 9], and increases to band gaps [10, 11]. These trends have been established in bulk materials from concurrent studies of bonding environments, obtained from crystal structure refinements of diffraction data, and measurements of electronic, magnetic, and optical properties.

Epitaxial heterostructures present unique opportunities for controlling octahedral rotations and enabling novel strategies to exploit rotations for new functionalities [12, 13]. While changes to material composition are needed to tune rotations in bulk perovskites, thin films enable two approaches to control the octahedral network independent of composition. The first approach is epitaxial strain [14], in which the film takes on the in-plane lattice parameters of the substrates below a critical thickness for strain relaxation. In strained perovskites, the octahedral rotations will be modified to accommodate these new lattice parameters, often through distinct changes to in-plane and out-of-plane rotations leading to increased anisotropy in the bonding environment compared to their bulk counterparts [15, 16]. The second approach is interfacial octahedral coupling, in which the rotations in the near-interfacial region are modified from the non-interfacial region due to the geometric constraints of connecting distinct rotational patterns and/or magnitudes at the heterojunction [17–19]. This structural coupling occurs over a much smaller length scale, often only a few unit cells [20], compared to strain-based effects that can be stabilized for tens of nms. This ability to locally manipulate octahedral rotations via interfaces is a particularly powerful approach for advancing fundamental understanding of perovskite physics and for realizing new physical phenomena not present in bulk materials. A non-comprehensive list of property design based on interfacial octahedral coupling includes improper multiferroics [21–23], spatially modulated exchange interactions [24], local engineering of metal–insulator transitions [25, 26], control of magnetocrystalline anisotropy [27, 28], realization of polar metals [29, 30], and structural delta-doping [31].

The flipside to these new scientific opportunities is the general challenge of measuring octahedral rotations in epitaxial thin films and superlattices. This material geometry precludes the use of X-ray or neutron powder diffraction, approaches commonly used for bulk materials. Instead, researchers must turn to synchrotron X-ray scattering or scanning transmission electron microscopy. This chapter aims to provide a tutorial guide for the most commonly employed X-ray-based approach for gaining insights into octahedral rotations: the measurement and analysis of half-order diffraction peaks. This technique provides an average value of the oxygen positions throughout the film and therefore enables a refinement of the crystal structure of strained films, which is not possible for perovskite oxides simply by measuring the integer Bragg reflections. This approach does not provide spatially resolved information, such as how the rotations are perturbed at interfaces. Layer-resolved rotational information can be obtained from coherent Bragg rod analysis (COBRA) [32–34], an advanced synchrotron scattering approach that requires ultrathin films typically less than 15-unit cells.

2 Determining the Rotation Pattern

2.1 Measurement Approach

The measurement of octahedral rotations using X-ray scattering is based on the simple idea that the oxygen displacements effectively double the size of the pseudocubic unit cell, thereby creating reciprocal lattice points at half-integer values of H, K, and L. Figure 2 illustrates this connection between octahedral rotations in real space and their implications for half-order diffraction peaks in reciprocal space. As pointed out by Glazer in 1975 [35], each type of rotation—for instance, a^- or b^+—results in a distinct set of half-order diffraction peaks. Assuming there are no symmetry lowering distortions associated with the *A*-site or *B*-site cations, measurement of these half-order peaks provides a direct fingerprint of the rotation pattern within a perovskite. Table 1 lists the half-order reflections produced by each rotation.

This connection between a rotation type and Bragg peaks enables the rotation pattern within a thin film to be determined through systemically surveying the presence and absence of half-order peaks. For instance, the common $a^-a^-c^+$ (*Pbnm*) rotation pattern can be identified by establishing the presence of an in-phase and two out-of-phase rotations. The presence of a $\frac{1}{2}\,\frac{3}{2}\,1$ reflection and the absence of $\frac{1}{2}\,1\,\frac{3}{2}$ and $1\,\frac{1}{2}\,\frac{3}{2}$ peaks indicate in-phase rotations about the *c*-axis, typically denoted as the out-of-plane direction in (001)-oriented perovskite films. The presence of $\frac{3}{2}\,\frac{1}{2}\,\frac{1}{2}$, $\frac{1}{2}\,\frac{3}{2}\,\frac{1}{2}$, and $\frac{1}{2}\,\frac{1}{2}\,\frac{3}{2}$ peaks is indicative of out-of-phase rotations within the in-plane directions, i.e. the pseudocubic [100] and [010] directions.

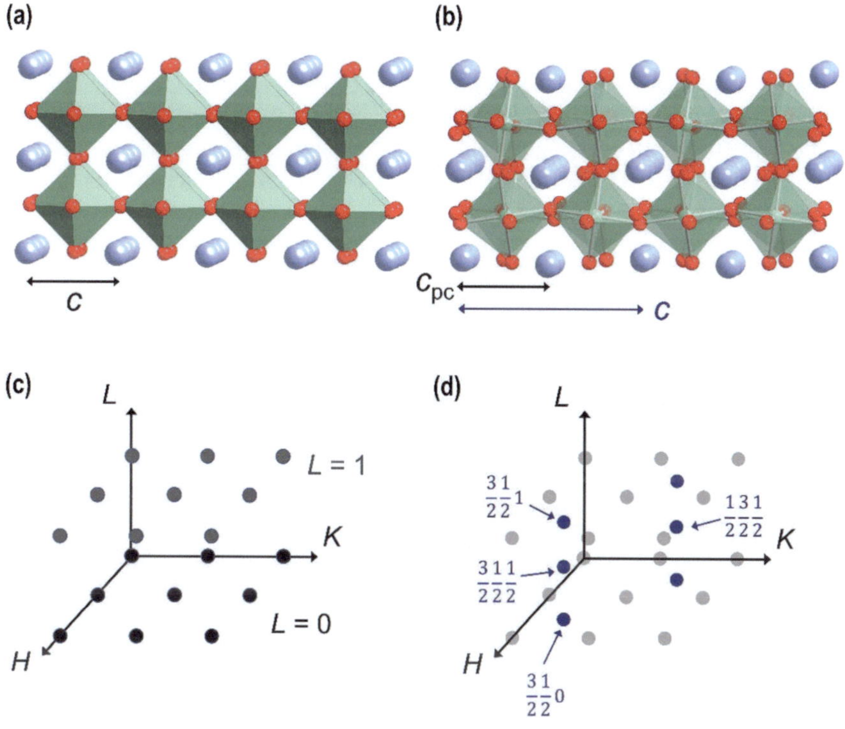

Fig. 2 Schematic relating the real and reciprocal space lattices of perovskites. The crystal structures of **a** a cubic ($a^0a^0a^0$) perovskite without rotations and **b** an orthorhombic perovskite with $a^-a^-c^+$ rotations. The rotations effectively double the unit cell, resulting in half-order diffraction peaks. Schematics of the reciprocal lattice in an **c** $a^0a^0a^0$ and **d** $a^-a^-c^+$ perovskite. The half-order reflections are given by the blue points with the integer reflections lightened for clarity. In **d**, octahedral rotations are assumed to be the only contributions to the half-order peaks; for simplicity, the effects of *A*-site displacements that occur in orthorhombic perovskites are not included

Table 1 A list of the half-order diffraction conditions that arise from octahedral rotations

Rotation type	Result peaks	Example peaks
a^-	$\frac{h}{2}\frac{k}{2}\frac{l}{2}$, where h, k, l are all odd and $k \neq l$	$\frac{1}{2}\frac{1}{2}\frac{3}{2}$, $\frac{1}{2}\frac{3}{2}\frac{1}{2}$
b^-	$\frac{h}{2}\frac{k}{2}\frac{l}{2}$, where h, k, l are all odd and $h \neq l$	$\frac{1}{2}\frac{1}{2}\frac{3}{2}$, $\frac{3}{2}\frac{1}{2}\frac{1}{2}$
c^-	$\frac{h}{2}\frac{k}{2}\frac{l}{2}$, where h, k, l are all odd and $h \neq k$	$\frac{3}{2}\frac{1}{2}\frac{1}{2}$, $\frac{1}{2}\frac{3}{2}\frac{1}{2}$
a^+	$h\frac{k}{2}\frac{l}{2}$, where k and l are odd, h is even or odd, and $k \neq l$	$1\frac{1}{2}\frac{3}{2}$, $2\frac{3}{2}\frac{5}{2}$
b^+	$\frac{h}{2}k\frac{l}{2}$, where h and l are odd, k is even or odd, and $h \neq l$	$\frac{1}{2}1\frac{3}{2}$, $\frac{3}{2}2\frac{5}{2}$
c^+	$\frac{h}{2}\frac{k}{2}l$, where h and k are odd, l is even or odd, and $h \neq k$	$\frac{1}{2}\frac{3}{2}1$, $\frac{3}{2}\frac{5}{2}2$

However, there are practical issues that often must be considered and planned for in order to draw solid conclusions from measurements of half-order diffraction peaks.

1. Peak intensities: Is there enough intensity to measure a peak? Compared to integer peaks such as 002 where the *A*- and *B*-site cations contribute to scattered intensity, half-order reflections are typically 2–3 orders of magnitude weaker. For films that are 10s of nm or less in thickness, this likely requires a synchrotron source to measure the half-order diffraction peaks. In some perovskite structures, most notably the $a^-a^-c^+$ *Pbnm* structure, the oxygen displacements are accompanied by antiferroic *A*-site displacements that also contribute to half-order peak intensities. These cation displacements simplify measurement of certain half-order peaks by increasing their magnitude, but also complicate quantitative analysis of peak intensities as will be discussed in the next section.
2. Rotational magnitudes: Can one distinguish between $a^-b^-c^+$ and $a^-a^-c^+$? The presence or absence of a diffraction peak enables identification of + or − type rotations along different axes, but determining the rotational magnitude requires quantitative analysis of multiple peak intensities. Thus, the presence of $\frac{1}{2}\frac{3}{2}1$, $\frac{3}{2}\frac{1}{2}\frac{1}{2}$, $\frac{1}{2}\frac{3}{2}\frac{1}{2}$, and $\frac{1}{2}\frac{1}{2}\frac{3}{2}$ peaks is fully consistent with the $a^-a^-c^+$ rotation pattern, but this collection of peaks is also consistent with a $a^-b^-c^+$ pattern. Other physical arguments may be used to support assigning the structure to $a^-a^-c^+$ or $a^-b^-c^+$. For instance, if the film is strained to a cubic substrate, then the two in-plane bond angles should be equivalent, requiring an $a^-a^-c^+$ structure. If grown on a lower symmetry substrate with different in-plane lattice parameters, such as orthorhombic $DyScO_3$ or $NdGaO_3$, then it is unlikely that the bonding geometries along the pseudocubic *a*- and *b*-axes will be identical, leading to an $a^-b^-c^+$ structure.
3. Peak degeneracy: How to distinguish between b^- and c^- rotations if they both produce a $\frac{3}{2}\frac{1}{2}\frac{1}{2}$ reflection? Unlike the in-phase rotations, the out-of-phase rotations along a single axis cannot be isolated and conclusively identified from measurement of a single peak as there is overlap in the resultant peaks from different rotation axes. As the above question illustrates, the presence of a $\frac{3}{2}\frac{1}{2}\frac{1}{2}$ could indicate a b^- rotation, a c^- rotation, or the existence of both rotations

within a sample. Here again some basic symmetry arguments can help to disentangle these contributions as the two in-plane rotations (a and b) are likely to be similar in magnitude, so it is highly unlikely that a film will have a $a^-b^0c^-$ or similar pattern. Additionally, tensile strain tends to decrease the magnitude of the c^- or c^+ rotations, while compressive strain increases these rotations. These strain dependences can help guide interpretation of families of half-order peaks such as the trio of $\frac{3}{2}\frac{1}{2}\frac{1}{2}$, $\frac{1}{2}\frac{3}{2}\frac{1}{2}$, and $\frac{1}{2}\frac{1}{2}\frac{3}{2}$. However, conclusively establishing the difference between a c^- rotation with a very small magnitude and the lack of a rotation (c^0) requires quantitative peak intensity analysis if there are a^-a^- in-plane rotations.

4. Orientational domains: Is this film $a^+b^+c^-$ or are there domains of both $a^-b^+c^-$/$a^+b^-c^-$? This question might arise if a survey scan is carried out on a film yielding peaks at $1\,\frac{1}{2}\,\frac{3}{2}$ and $\frac{1}{2}\,1\,\frac{3}{2}$, but not at $\frac{1}{2}\,\frac{3}{2}\,1$. One might be tempted to conclude an $a^+b^+c^-$ structure. But this rotation pattern is quite rare in perovskite oxides, especially compared to the common $a^-a^-c^+$ (or $a^-b^+a^-$) pattern. Measurement of $\frac{h}{2}\frac{k}{2}\frac{l}{2}$ type peaks can clarify the situation. In particular, peak intensity at the $\frac{1}{2}\frac{1}{2}\frac{3}{2}$ is indicative of an a^- or b^- rotation, which is incompatible with $a^+b^+c^-$, pointing to the presence of spatially inhomogeneous distribution of the in-phase axis along both the a^- and b^- in-plane axes. These orientational domains, illustrated in Fig. 3a, have been measured in many *Pbnm*-type thin films grown on cubic ($SrTiO_3$ or LSAT) or rhombohedral substrates ($LaAlO_3$) [36–43].

2.2 Examples: Insights from Peak Measurements

The measurement of half-order peaks, even without the quantitative analysis described in the next section, is a useful tool for understanding epitaxy, temperature-driven phase transitions, and the propensity for specific perovskite structural variants as a function of thickness or deposition conditions in oxide heterostructures. A few examples are given below to illustrate what kind of studies can be performed through just the measurement of half-order peaks.

Epitaxial orientations. The *Pbnm*-type structure ($a^-a^-c^+$) is one of the most common structural variants for ABO_3 perovskites, including many 3*d*, 4*d*, and 5*d* systems that are grown as epitaxial films. In these heterostructures, determination of the epitaxial relationship between the film and substrate requires measurement of the in-phase axis. For $(001)_{pc}$-oriented interfaces, there are three orientations: $a^+a^-c^-$, $a^-a^+c^-$ and $a^-a^-c^+$. Measurement of half-order diffraction peaks provides a straightforward means to distinguish between these orientations.

There are two principal energies that play a role in establishing if the in-phase axis is along the growth direction ($a^-a^-c^+$) or within the plane ($a^+a^-c^-$, $a^-a^+c^-$) of a (001)-oriented film. Typically in *Pbnm* perovskites, the pseudocubic unit cell is shorter along the in-phase axis compared to the out-of-phase axes. Therefore, the biaxial strain imposed from a substrate will favor the $a^-a^-c^+$ pattern under tensile

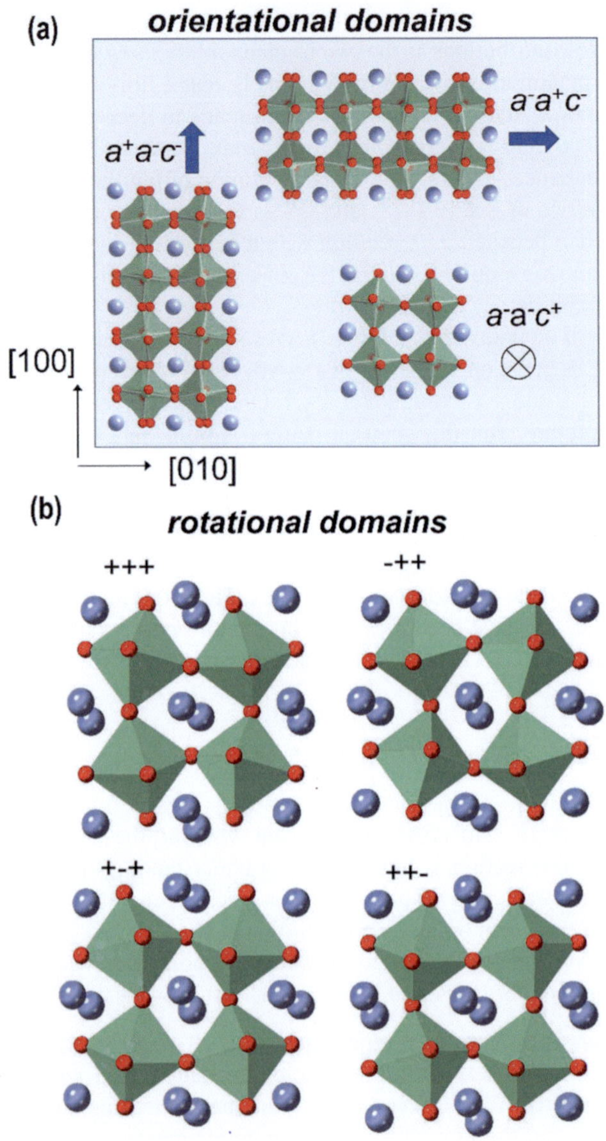

Fig. 3 Schematics contrasting orientational and rotational domains as illustrated with the $a^-a^-c^+$ (*Pbnm*) rotation pattern. **a** The three possible orientational domains for films the $a^-a^-c^+$ pattern grown on a pseudocubic (001)-oriented substrate. The in-phase axis rotation can lay along either of the two in-plane directions or the out-of-plane direction. **b** Four rotational domains created by applying either clockwise or counterclockwise rotations to the first octahedron along the three rotation axes. These patterns are created by using combinations of positive and negative values of α, β, and γ, all within the same $a^-a^-c^+$ pattern

strain and the $a^+a^-c^-/a^-a^+c^-$ patterns under compressive strain. The second important factor is the interfacial energy at the film/substrate junction, which will depend on the coupling of octahedral rotations. This energy plays a significant role in the growth of *Pbnm* films on *Pbnm* substrates such as $DyScO_3$, $GdScO_3$ or $NdGaO_3$. These substrates can be purchased in different orientations, leading to the in-phase axis within the plane [the case for (110)-oriented orthorhombic substrates] or out-of-the plane [the case for (001)-oriented orthorhombic substrates]. This interfacial coupling commonly overrides strain considerations and can produce films under compressive strain with $a^-a^-c^+$ patterns or tensile-strained films with $a^+a^-c^-$ patterns. The symmetry of the substrate also can play a determining role in the formation of orientational domains within the film. In heterostructures grown on cubic ($a^0a^0a^0$) or rhombohedral ($a^-a^-a^-$) substrates, films often exhibit a mixture of $a^+a^-c^-$ and $a^-a^+c^-$ orientations as the equivalent in-plane lattice parameters of the substrate lead to an energetic degeneracy of the two orientations and thus an equal chance of nucleating these structures during the initial stages of film growth. In contrast, *Pbnm*-type films grown on (110)-oriented orthorhombic substrates typically exhibit a single $a^+a^-c^-$ orientation.

These trends were illustrated by Choquette and coauthors who measured the rotation pattern for $EuFeO_3$ films and other *Pbnm* perovskites on a variety of substrates [39]. Compiling results from 15 different material systems grown on cubic substrates (and therefore lacking any substrate-imprinting of the in-phase axis), it was observed that mixed $a^+a^-c^-$ and $a^-a^+c^-$ structures were found under compressive strain while $a^-a^-c^+$ structures were stabilized under tensile strain larger than 2%. In samples under more modest tensile strain, all three orientations were observed including some materials that exhibited all three orientational variants within the same film. Analysis of intensities from the $1\,\frac{1}{2}\,\frac{3}{2}$, $\frac{1}{2}\,1\,\frac{3}{2}$, and $\frac{1}{2}\,\frac{3}{2}\,1$ families of peaks enabled the volume fraction of $a^-a^-c^+$ and $a^+a^-c^-/a^-a^+c^-$ structures to be quantified in films with mixed in-plane and out-of-plane orthorhombic orientations. In contrast, all films grown on (110)-oriented orthorhombic substrates exhibit single $a^+a^-c^-$ (or $a^+b^-c^-$) orientations with the in-phase axis imprinted from the substrate even for films under $> 2\%$ tensile strain. This interfacial coupling between an orthorhombic substrate and film has been observed using half-order diffraction measurements for numerous materials systems, including $CaTiO_3/NdGaO_3$ [44], $EuFeO_3/DyScO_3$ [39], $SrRuO_3/GdScO_3$ [45], $Ca_{0.5}Sr_{0.5}IrO_3/GdScO_3$ [46], and $La(Cr_{0.2}Mn_{0.2}Fe_{0.2}Co_{0.2}Ni_{0.2})O_3/NdGaO_3$ [47]. An interesting exception is $SrRuO_3$ grown under compressive strain on (110)-oriented $NdGaO_3$, in which case the $a^-a^+c^-$ pattern in bulk $SrRuO_3$ is converted to $a^0a^0c^-$ due to the compressive strain, which reduces the α and β magnitudes to 0° thus removing the in-phase axis [48].

A *Pbnm* substrate can even imprint an in-phase rotation into a thin film that lacks such a rotation in bulk. For example, Chen and coauthors recently used half-order diffraction to show that a 9-unit cell thick $La_{0.67}Sr_{0.33}MnO_3$ film on $NdGaO_3$ hosted an $a^+a^-c^-$ rotation pattern [49], despite that bulk $La_{0.67}Sr_{0.33}MnO_3$ crystallizes in an $a^-a^-a^-$ structure. In contrast, a thicker 110-unit cell $La_{0.67}Sr_{0.33}MnO_3$ film was shown to revert to out-of-phase rotations about all three axes. This study is also notable for demonstrating how to correlate integer peak positions in reciprocal space maps with rotational patterns determined from half-order reflections.

Phase transitions. Measurement of rotation-based diffraction peaks provides a direct way to study temperature-dependent phase transitions between structural variants. Studies of $SrTiO_3$ films offer clear examples of this approach. $SrTiO_3$ exhibits a canonical displacive phase transition at 105 K as it goes from a cubic structure ($a^0a^0a^0$) to a tetragonal structure ($a^0a^0c^-$) below the transition. The tetragonal phase is accompanied by the presence of $\frac{3}{2}\frac{1}{2}\frac{1}{2}$ and $\frac{1}{2}\frac{3}{2}\frac{1}{2}$ type peaks arising from the c^- rotation. Tracking the temperature dependence of half-order peaks in $SrTiO_3$ films grown on different substrates and thicknesses, He and coauthors were able to show a dramatic (modest) increase in the transition temperature for $SrTiO_3$ under compressive (tensile) strain [50, 51]. By measuring multiple samples at different strain states, they demonstrated that the phase transition can vary from 125 to 330 K due to epitaxial constraints.

Similar studies have been performed on other epitaxial perovskites to illuminate phase transitions in ultrathin films. In $LaAlO_3$ films grown on $SrTiO_3$, the transition from the *P4/mmm* phase that lacks rotations to the *Imma* structure that hosts rotations was shown to occur near 970 K in a 24-unit cell thick film by tracking the $\frac{3}{2}\frac{1}{2}\frac{1}{2}$ peak intensity [32]. In contrast, a 9-unit cell thick film undergoes the same transition at 813 K in close agreement with bulk $LaAlO_3$. The changes in transition temperature were attributed to differences in unit cell volume of the films at different thicknesses. A substrate dependent phase transition was identified in $Sr_{0.5}Ca_{0.5}IrO_3$, where films on $SrTiO_3$ transition into an $a^+a^-a^0$ phase near 510 K while films on $GdScO_3$ substrates retain their $a^+b^-c^-$ structure from room temperature to 520 K [46]. Films of $AgTa_{0.5}Nb_{0.5}O_3$, which hosts numerous structural phases including one that exhibits a complex 4-unit cell long rotation pattern, were studied through the measurement of satellite peaks off of the $\frac{3}{2}\frac{1}{2}\frac{1}{2}$ reflection as well as a collection of other half-order peaks [52, 53]. Both compressive and tensile strain were demonstrated to have significant impacts on the phase transition temperature, in some cases altering the transitions by > 250 K.

Structural impact on properties. Determining the rotation pattern and crystal symmetry, even without refining atomic positions, can provide important insights into the origin of physical properties in films. While bulk $SrRuO_3$ is orthorhombic (*Pbnm*) at room temperature and transitions to a tetragonal (*I4/mcm*) phase at 820 K, in thin films both an orthorhombic-like ($a^+a^-c^-$ family of rotation patterns) and the tetragonal ($a^0a^0c^-$) phase can be stabilized. The structures exhibit significant differences in magnetic anisotropy, magnetotransport phenomena, and Curie temperature [54–56]. Measurement of the presence or absence of the $1\,\frac{1}{2}\,\frac{3}{2}$ family of peaks that accompanies the in-phase rotation axis as well as the $\frac{1}{2}\frac{1}{2}\frac{3}{2}$ peak arising from a^- has been used to demonstrate how the phase stability depends on film thickness [57], deposition conditions [58], strain [48], and can even be modified via helium ion implantation [59]. This understanding can then be used to assist interpretation of and assign physical mechanisms to differences in magnetocrystalline anisotropy or

Hall effects between $SrRuO_3$ films differing in thickness, substrate, or the presence of buffer layers [27, 60, 61]. Additionally, diffraction measurements to determine if films have uniform or mixed in-phase axis orientation have revealed the impact of structural domains on magnetic behavior in $SrRuO_3$ [40].

Grutter and coauthors used half-order diffraction characterization to gain insight into the magnetic properties of $CaMnO_3/CaRuO_3$ superlattices [62]. Magnetometry experiments revealed an oscillatory pattern in their saturation magnetization, altering from a large to small value when the number of $CaMnO_3$ unit cells within each superlattice period was altered from even to odd. Through measurement of $1\,1\,\frac{3}{2}$ and $0\,\frac{1}{2}\,\frac{3}{2}$ reflections from superlattices with different layer parity, the authors were able to conclude that the interfacial symmetry-matching and presence of orthorhombic domains were key features in the magnitude of the magnetization. In $(SrIrO_3)_n/(SrTiO_3)_1$ superlattices, a relative suppression of the α and β rotations in an $n = 1$ structure compared to an $n = 2$ structure was identified from half-order diffraction peaks and linked to differences in magnetic interactions between the two superlattices [63].

Beyond just measuring the presence or absence of peaks, straightforward analysis of peak widths also provides an additional avenue for material characterization. Quantitative information about orientational domain sizes (ξ) can be extracted from the full-width at half maximum (FWHM) of half-order peaks scanned through H or K using Scherrer analysis, where $\xi \approx \frac{2\pi}{\beta}$ and β is the FWHM in Å^{-1}. This analysis was used by de Andrés et al. to examine the relationship between ξ, film thickness, and magnetic properties of (001)-oriented $La_{0.7}Ca_{0.3}MnO_3$ films [64].

3 Quantifying Oxygen Positions and Bonding Geometry

3.1 Measurement and Analysis Approach

Refining oxygen positions. The oxygen positions, and thus the rotation and bond angles, can be determined through quantitative analysis of the half-order peak intensities [16]. This approach is based on the measurement on numerous families of half-order reflections, from which their intensities are extracted. The intensities of these same peaks are then calculated using the rotational angles as free parameters. By minimizing the difference between the experimental and simulated peak intensities, the positions of the oxygen atoms are obtained.

The diffraction peak intensities are calculated through the structure factor (F_{hkl}),

$$F_{hkl} = \sum_{n}^{N} f_n e^{2\pi i(hu+kv+lw)} \tag{1}$$

Here, N is the number of unique atoms within the unit cell; f_n is the atomic form factor of the nth atom; u, v, and w are the atomic point coordinates; and h, k, and l are the Miller indices. Doubling the pseudocubic perovskite unit cell along each Cartesian direction results in 24 oxygen atoms, 8 A-site cations, and 8 B-site cations, resulting in the following structure factor used in half-order analysis,

$$F_{hkl} = f_{O^{2-}} \sum_{n=1}^{24} f_n e^{2\pi i(hu+kv+lw)} + f_A \sum_{n=1}^{8} f_n e^{2\pi i(hu+kv+lw)} + f_B \sum_{n=1}^{8} f_n e^{2\pi i(hu+kv+lw)} \quad (2)$$

In the structure factor calculated using this new doubled unit cell, the indices of the Bragg reflections must also be doubled. This means that a half-order index, with respect to the 5-atom pseudocubic perovskite cell, is entered into Eq. 2 as an odd integer. For instance, the structure factor for a 0 0 ½ peak is calculated by entering $h = 0$, $k = 0$, and $l = 1$ into Eq. 2. The atomic form factors used for Eq. 2 should be the values of the atoms in their anionic or cationic charge states, for instance O^{2-} or Sr^{2+}. Values of $f_{O^{2-}}$ are reported in Ref. [65], while form factors for different cations are found in Ref. [66].

The oxygen positions are determined by the magnitude of the rotations. Following Glazer [1], α, β, and γ correspond to the rotational magnitudes along the pseudocubic [100], [010], and [001] directions, as shown in Fig. 1. Note that these angles are distinct from the B–O–B bond angles (θ), as the bond angles result from the combined effects of α, β, and γ. For instance, $\theta_{ab} \sim 180° - 2\sqrt{\alpha^2 + \gamma^2}$ and $\theta_c \sim 180° - 2\sqrt{2\alpha^2}$ for small values of α and γ (~ < 10°), where θ_{ab} and θ_c are the in-plane and out-of-plane bond angles, respectively. Tables 2 and 3 list the oxygen positions as a function of α, β, and γ for the $a^-a^-c^-$ and $a^-a^-c^+$ rotation patterns, demonstrating how the rotations displace the anions. The differing c^- and c^+ rotations alter oxygen positions in the upper layer of octahedra ($w \geq 0.5$), changing the sign of the tan (γ) terms due to the difference in phase of the rotation. The $a^+a^-c^-$ positions can be obtained by switching the u and w point coordinates, exchanging α and β, as well as a and c [67]. It should be noted that these positions are obtained by applying the rotations about the original (non-displaced) Cartesian axes, which can lead to octahedral distortions as the rotations become large in magnitude (> 10°). Fowlie and coauthors demonstrated an iterative approach for incrementally applying the rotations sequentially about the three axes that is effective in mitigating this problem [68].

For perovskites where the anions are the only atoms displaced from the ideal cubic positions, the f_A and f_B summations will be zero for half-order peaks, and thus only the oxygen atoms contribute to the half-order intensity. This scenario is found in the perovskite structures $a^-a^-a^-$ ($R\bar{3}c$, for example bulk $LaNiO_3$ and $La_{0.7}Sr_{0.3}MnO_3$), $a^-a^-c^-$ ($C2/c$, biaxially strained $LaNiO_3$ and $La_{0.7}Sr_{0.3}MnO_3$), and $a^0a^0c^-$ ($I4/mcm$, for example, low temperature $SrTiO_3$). In this case, the structure factor for the half-order reflections will contain 24 terms within the summation. For perovskites

Table 2 Oxygen atomic point coordinates as a function of the α, β, and γ rotation angles for the $a^-a^-c^-$ pattern

Atom	u	v	w
O1	$0.25 + c/(4a)\tan(\beta)$	$0.25 - c/(4b)\tan(\alpha)$	0
O2	$0.25 - c/(4a)\tan(\beta)$	$0.75 + c/(4b)\tan(\alpha)$	0
O3	$0.75 + c/(4a)\tan(\beta)$	$0.75 - c/(4b)\tan(\alpha)$	0
O4	$0.75 - c/(4a)\tan(\beta)$	$0.25 + c/(4b)\tan(\alpha)$	0
O5	0	$0.25 + a/(4b)\tan(\gamma)$	$0.25 - a/(4c)\tan(\beta)$
O6	$0.25 - b/(4a)\tan(\gamma)$	0	$0.25 + b/(4c)\tan(\alpha)$
O7	0.5	$0.25 - a/(4b)\tan(\gamma)$	$0.25 + a/(4c)\tan(\beta)$
O8	$0.25 + b/(4a)\tan(\gamma)$	0.5	$0.25 - b/(4c)\tan(\alpha)$
O9	0	$0.75 - a/(4b)\tan(\gamma)$	$0.25 + a/(4c)\tan(\beta)$
O10	0.5	$0.75 + a/(4b)\tan(\gamma)$	$0.25 - a/(4c)\tan(\beta)$
O11	$0.75 - b/(4a)\tan(\gamma)$	0.5	$0.25 + b/(4c)\tan(\alpha)$
O12	$0.75 + b/(4a)\tan(\gamma)$	0	$0.25 - b/(4c)\tan(\alpha)$
O13	$0.25 - c/(4a)\tan(\beta)$	$0.25 + c/(4b)\tan(\alpha)$	0.5
O14	$0.25 + c/(4a)\tan(\beta)$	$0.75 - c/(4b)\tan(\alpha)$	0.5
O15	$0.75 - c/(4a)\tan(\beta)$	$0.75 + c/(4b)\tan(\alpha)$	0.5
O16	$0.75 + c/(4a)\tan(\beta)$	$0.25 - c/(4b)\tan(\alpha)$	0.5
O17	0	$0.25 - a/(4b)\tan(\gamma)$	$0.75 + a/(4c)\tan(\beta)$
O18	$0.25 + b/(4a)\tan(\gamma)$	0	$0.75 - b/(4c)\tan(\alpha)$
O19	0.5	$0.25 + a/(4b)\tan(\gamma)$	$0.75 - a/(4c)\tan(\beta)$
O20	$0.25 - b/(4a)\tan(\gamma)$	0.5	$0.75 + b/(4c)\tan(\alpha)$
O21	0	$0.75 + a/(4b)\tan(\gamma)$	$0.75 - a/(4c)\tan(\beta)$
O22	0.5	$0.75 - a/(4b)\tan(\gamma)$	$0.75 + a/(4c)\tan(\beta)$
O23	$0.75 + b/(4a)\tan(\gamma)$	0.5	$0.75 - b/(4c)\tan(\alpha)$
O24	$0.75 - b/(4a)\tan(\gamma)$	0	$0.75 + b/(4c)\tan(\alpha)$

Here the c-axis is taken as the out-of-plane (growth) direction for the film, while a and c are the in-plane and out-of-plane pseudocubic lattice parameters

with A-site displacements, as will be discussed later, the structure factor needed to calculate half-order intensities must also include 8 additional terms for the A-site cations.

From the structure factor, the peak intensities (I_{hkl}) can be calculated using,

$$I_{hkl} = I_0 G \sum_{j=1}^{4} D_j F_{hkl} F_{hkl}^* \tag{3}$$

where F_{hkl}^* is the complex conjugate of the structure factor, I_0 is the incident beam intensity, G accounts for geometric factors, and D_j indicates the volume fraction

Table 3 Oxygen atomic point coordinates as a function of the α, β, and γ rotation angles for the $a^-a^-c^+$ pattern

Atom	u	v	w
O1	$0.25 + c/(4a)\tan(\beta)$	$0.25 - c/(4b)\tan(\alpha)$	0
O2	$0.25 - c/(4a)\tan(\beta)$	$0.75 + c/(4b)\tan(\alpha)$	0
O3	$0.75 + c/(4a)\tan(\beta)$	$0.75 - c/(4b)\tan(\alpha)$	0
O4	$0.75 - c/(4a)\tan(\beta)$	$0.25 + c/(4b)\tan(\alpha)$	0
O5	0	$0.25 + a/(4b)\tan(\gamma)$	$0.25 - a/(4c)\tan(\beta)$
O6	$0.25 - b/(4a)\tan(\gamma)$	0	$0.25 + b/(4c)\tan(\alpha)$
O7	0.5	$0.25 - a/(4b)\tan(\gamma)$	$0.25 + a/(4c)\tan(\beta)$
O8	$0.25 + b/(4a)\tan(\gamma)$	0.5	$0.25 - b/(4c)\tan(\alpha)$
O9	0	$0.75 - a/(4b)\tan(\gamma)$	$0.25 + a/(4c)\tan(\beta)$
O10	0.5	$0.75 + a/(4b)\tan(\gamma)$	$0.25 - a/(4c)\tan(\beta)$
O11	$0.75 - b/(4a)\tan(\gamma)$	0.5	$0.25 + b/(4c)\tan(\alpha)$
O12	$0.75 + b/(4a)\tan(\gamma)$	0	$0.25 - b/(4c)\tan(\alpha)$
O13	$0.25 - c/(4a)\tan(\beta)$	$0.25 + c/(4b)\tan(\alpha)$	0.5
O14	$0.25 + c/(4a)\tan(\beta)$	$0.75 - c/(4b)\tan(\alpha)$	0.5
O15	$0.75 - c/(4a)\tan(\beta)$	$0.75 + c/(4b)\tan(\alpha)$	0.5
O16	$0.75 + c/(4a)\tan(\beta)$	$0.25 - c/(4b)\tan(\alpha)$	0.5
O17	0	$0.25 + a/(4b)\tan(\gamma)$	$0.75 + a/(4c)\tan(\beta)$
O18	$0.25 - b/(4a)\tan(\gamma)$	0	$0.75 - b/(4c)\tan(\alpha)$
O19	0.5	$0.25 - a/(4b)\tan(\gamma)$	$0.75 - a/(4c)\tan(\beta)$
O20	$0.25 + b/(4a)\tan(\gamma)$	0.5	$0.75 + b/(4c)\tan(\alpha)$
O21	0	$0.75 - a/(4b)\tan(\gamma)$	$0.75 - a/(4c)\tan(\beta)$
O22	0.5	$0.75 + a/(4b)\tan(\gamma)$	$0.75 + a/(4c)\tan(\beta)$
O23	$0.75 - b/(4a)\tan(\gamma)$	0.5	$0.75 - b/(4c)\tan(\alpha)$
O24	$0.75 + b/(4a)\tan(\gamma)$	0	$0.75 + b/(4c)\tan(\alpha)$

Here the c-axis is taken as the out-of-plane (growth) direction for the film, while a and c are the in-plane and out-of-plane pseudocubic lattice parameters

of different rotational domains. The geometric prefactor depends on the beam footprint, which goes as $1/\sin(\eta)$ where η is the incident beam angle, and the Lorentz polarization correction. At many synchrotron facilities, data is already corrected for these geometric terms and the beamline scientist will have information about relevant corrections needed for data analysis. An additional Debye–Waller term ($e^{-B_n\left(\frac{q}{4\pi}\right)^2}$ where B_n is the Debye–Waller factor) can also be included in Eq. 3 [32]. Inclusion of this q-dependent term that accounts for thermal motion can become especially important when trying to fit peak intensities for reflections with very different q magnitudes.

The rotational domains accounted for through D_j do not refer to spatial differences in the rotation pattern such as a mixed $a^-a^+c^-/a^+a^-c^-$ structure, but instead capture structural differences that arise from whether the octahedron closest to the origin rotates clockwise or counterclockwise about each axis. Put differently, these domains arise from combinations of positive and negative values of α, β, and γ within the same rotational pattern (Fig. 3b). There are four domains that result in differing half-order intensities: all α, β, and γ values are positive (+++) and two of these magnitudes are positive (−++, +−+, and ++−). Each of the other four possible combinations of positive and negative values result in intensities that are identical with one of the +++, −++, +−+, and ++− domains and thus do not need to be considered.

The values of α, β, γ, and D_j are then used as fitting parameters to minimize the squared difference between the experimental and calculated intensities (= $\sum_n \left(I_{calc} - I_{\exp}\right)_n^2$ or = $\sum_n \frac{\left(I_{calc}-I_{\exp}\right)_n^2}{I_{\exp,n}}$, over the n measured peaks). Prior to running the fitting procedure, the intensities are normalized either to the most intense half-order peak or to the 002 or 004 reflection. Note that these integer reflections contain contributions from the *B*-site cations and therefore the third term on the right side of Eq. 2 must be included in the intensity calculations of these peaks. To simplify the fits, the domain volumes are typically determined directly from the measurement of symmetrically equivalent families of peaks, such as the $\pm\frac{1}{2}\ \pm\frac{1}{2}\ \frac{3}{2}$ reflections [16, 47]. An equal intensity among these peaks indicates an equal population of these domains. This is frequently observed in films grown on cubic substrates such as $SrTiO_3$ ($a^0a^0a^0$), which do not imprint rotations into a film. For such films, the four D_j terms are set to 0.25. In contrast, substrates that do have rotations, such as rhombohedral $LaAlO_3$ ($a^-a^-a^-$) or orthorhombic $NdGaO_3$ ($a^-a^-c^+$) have been shown to imprint rotational domain populations into epitaxial films [39]. In such cases, quantitative analysis of families of half-order peaks are required to determine the four D_j values. In addition to imprinting domains, substrates can impact quantitative analysis by themselves hosting strong and sharp half-order reflections as is the case for many commonly used non-cubic perovskites substrates such as aluminates, scandates, and gallates. In such situations, a simple Gaussian fit with an appropriate background term can be applied to the film half-order reflections to separate intensity from the film and substrate. While $(La,Sr)(Al,Ta)O_3$ (LSAT) is a commonly used substrate, synchrotron measurement and analysis of half-order peaks from films grown on this material is challenging as the substrate exhibits strong and very broad half-order peaks despite being a cubic ($a^0a^0a^0$) perovskite. Due to the large width of the half-order peaks of LSAT, measuring weak half-order peaks from thin films on this substrate is often not possible.

A-site displacements. Some perovskite structures, mostly notably the *Pbnm* $a^-a^-c^+$ phase shown in Fig. 3, also exhibit antiferroic *A*-site displacements. Like the oxygen displacements, these cation displacements are unit cell doubling and therefore contribute intensity to the half-order reflections. As a result, structural refinement of *Pbnm*-type perovskite films requires additional fitting parameters to capture the *A*-site positions [67], which are denoted as d_1 and d_2. Figure 4 displays the relationship of the *A*-site displacements to the octahedral tunnels created by the rotations and illustrates how these displacements are reproduced through the parameters d_1 and d_2. The d_1 displacement moves the *A*-site along the pseudocubic <110> family of directions, while the d_2 displacements act along the <100> family of directions. The atomic point coordinates of the *A*-site cations are given in Table 4 as a function of the displacement parameters. Similar to the rotational domains that arise from positive or negative values of α, β, and γ, there are four configurations of the *A*-site displacements within each rotational domain corresponding to positive and negative values of d_1 and d_2. These four configurations are denoted 1R, 2R, 1L, and 2L as shown in Fig. 4. Of these configurations, only two displace the *A*-site cations along the long diagonal of rhombus and thus are considered energetically favorable within a given rotational domain. In the case where all four rotational domains are present in equal populations, one then expects to find the four *A*-site displacement configurations present in equal populations and thus the four combinations can be weighed equally in the term accounting for *A*-site positions in Eq. 2 [47]. Measurements of symmetrically equivalent half-order peaks should be performed to confirm the equal population of the four *A*-site configurations. In *Pbnm* perovskites, fitting the intensities of half-order peaks using α, β, γ, d_1, d_2, and D_j thus yields the atomic positions for both the oxygen atoms and *A*-site cations enabling full structural refinement [41, 47, 67].

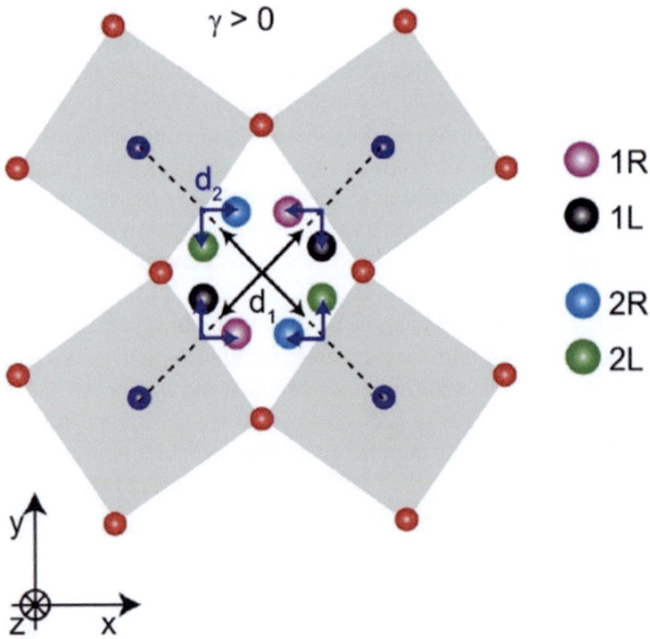

Fig. 4 Schematic illustrating the d_1 and d_2 *A*-site displacement parameters for a perovskite with an $a^-a^-c^+$ rotation pattern and $\gamma > 0$. Figure reproduced from Ref. [67]

Table 4 Atomic point coordinates for *A*-site cations as a function of the d_1 and d_2 parameters for the $a^-a^-c^+$ pattern

Atom	u	v	w
A1	d_1	$d_1 + d_2$	0
A2	$0.5 + d_1 + d_2$	d_1	0
A3	$0.5 + d_1$	$0.5 + d_1 + d_2$	0
A4	$d_1 + d_2$	$0.5 + d_1$	0
A5	$- d_1$	$- d_1 - d_2$	0.5
A6	$0.5 - d_1 - d_2$	$- d_1$	0.5
A7	$0.5 - d_1$	$0.5 - d_1 - d_2$	0.5
A8	$- d_1 - d_2$	$0.5 - d_1$	0.5

Here the *c*-axis is taken as the out-of-plane (growth) direction for the film

Obtaining the crystal structure. Once the atomic positions are obtained by minimizing the difference between the measured and calculated intensities, the full structural information can be presented with respect to the true unit cell (not the $2 \times 2 \times 2$ pseudocubic structure). The atomic point coordinates can be converted to positions in real space by multiplying the *u*, *v*, and *w* by the relevant doubled pseudocubic lattice parameter. From these positions, the in- and out-of-plane *B*–O–*B* bond angles and *B*–O bond lengths can be extracted. When comparing physical properties to structural parameters, it is these bond lengths and angles that are most directly coupled to electronic structure and magnetic interactions. The crystallographic unit cell and space group of the refined structure can be generated through crystallographic software. For example, the Bilbao Crystallographic Server (https://www.cryst.ehu.es/cgi-bin/cryst/programs/mcif2vesta/index.php) and the FINDSYM program (https://stokes.byu.edu/iso/findsym.php) can determine the space group and export a .cif file from an inputted list of atomic positions.

3.2 *Examples: Quantitative Peak Analysis*

Refining crystal structures in films. As described in the previous section, the general workflow for structural refinement of a perovskite oxide film is [67]:

(1) Measure the 002 or 004 reflection, which can be used for normalization of the half-order peaks as well as provide the *c*-axis parameter. The 113 or a similar peak should be measured from which the in-plane lattice parameter can be extracted.

(2) Survey the $1\,\frac{1}{2}\,\frac{3}{2}$, $\frac{1}{2}\,1\,\frac{3}{2}$, and $\frac{1}{2}\,\frac{3}{2}\,1$ half-order peaks to determine if the film hosts in-phase rotations. Similarly, measure the $\frac{1}{2}\,\frac{1}{2}\,\frac{3}{2}$, $\frac{1}{2}\,\frac{3}{2}\,\frac{1}{2}$, and $\frac{3}{2}\,\frac{1}{2}\,\frac{1}{2}$ peaks to confirm the presence of out-of-phase rotations.

(3) Having established the rotation pattern, a family of symmetrically equivalent peaks should be measured to determine if the rotational domain populations are equal. The group of $\frac{1}{2}\frac{1}{2}\frac{3}{2}$, $\frac{-1}{2}\frac{1}{2}\frac{3}{2}$, $\frac{1}{2}\frac{-1}{2}\frac{3}{2}$, and $\frac{-1}{2}\frac{-1}{2}\frac{3}{2}$ reflections is an example that can be used to determine the relative fraction of the domains.

(4) If the film has only out-of-phase rotations, then the $\frac{1}{2}\frac{1}{2}\frac{5}{2}$, $\frac{1}{2}\frac{5}{2}\frac{1}{2}$, and $\frac{5}{2}\frac{1}{2}\frac{1}{2}$ peaks should be measured. Along with the earlier measured $\frac{1}{2}\frac{1}{2}\frac{3}{2}$, $\frac{1}{2}\frac{3}{2}\frac{1}{2}$, and $\frac{3}{2}\frac{1}{2}\frac{1}{2}$ peaks, these reflections allow for quantitative analysis of α, β, and γ. A few other reflections with all half-order peaks can also be collected to improve accuracy of the refinement.

(5) If the film has in-phase rotations, then peaks with one integer and two half-order indices should be measured. For instance, if the film is $a^-a^+c^-$, then the $\frac{1}{2}\,1\,\frac{3}{2}$, $\frac{1}{2}\,1\,\frac{5}{2}$, and $\frac{1}{2}\,2\,\frac{3}{2}$ peaks should be collected to allow determination of β. As the presence of in-phase rotations is typically accompanied by *A*-site displacements, the $\frac{1}{2}\frac{1}{2}\frac{1}{2}$, $1\,\frac{3}{2}\,1$, and $\frac{3}{2}\frac{3}{2}\frac{3}{2}$ peaks should be measured (here again assuming $a^-a^+c^-$). These peaks will enable extraction of the d_1 and d_2 parameters.

An excellent example of this process comes from a recent study by Brahlek and coauthors, who measured oxygen and *A*-site displacements from a series of strained $La(Cr_{0.2}Mn_{0.2}Fe_{0.2}Co_{0.2}Ni_{0.2})O_3$ films on different substrates [47]. Figure 5 presents results from a film grown on $SrTiO_3$, which imposes 0.7% tensile strain. The authors first confirmed the rotation pattern of the film through measurements of the half-order peaks (Fig. 5a). Next, they confirmed that the films have an equal population of rotational domains (+++, +−+, etc.) and *A*-site displacement configurations (1R, 1L, etc.), as shown in Fig. 5b, followed by measurement of a number of half-order and integer peaks from which intensities were extracted. Figure 5c, d presents the evolution of the α, γ, d_1, and d_2 fitting parameters as the number of fitting iterations (n) progresses, until the fits converge on error-minimizing values of all parameters, approximately after 4 iterations. A comparison between the measured and simulated peak intensities is shown in Fig. 5e, which also provides a direct example of the collection of peaks needed to refine the atomic positions. Having established the *A*-site and oxygen coordinates, the crystal structure can be determined and visualized (Fig. 5f). In this specific material system, which is a high entropy oxide, the authors were interested if the chemical disorder on the *B*-site impacts the long-range crystalline order. Through the half-order diffraction analysis, the authors were able to demonstrate that the films retain macroscopic crystalline homogeneity with well-defined average values of metal–oxygen bond lengths and angles.

Structural impacts of strain, thickness, and superlattice formation. The use of synchrotron-based structural refinement has contributed significantly to the understanding of how the BO_6 bonding network in perovskite oxides is modified by epitaxial strain, decreased thickness, and the periodic interfaces formed in superlattices. As an example, we turn to studies of $LaNiO_3$-based heterostructures. The rare earth nickelates are one of the canonical families of perovskites oxides in which octahedral rotations play a central role in physical properties. The parent compound $LaNiO_3$ is a metal that crystallizes with an $a^-a^-a^-$ rotation pattern. Upon *A*-site

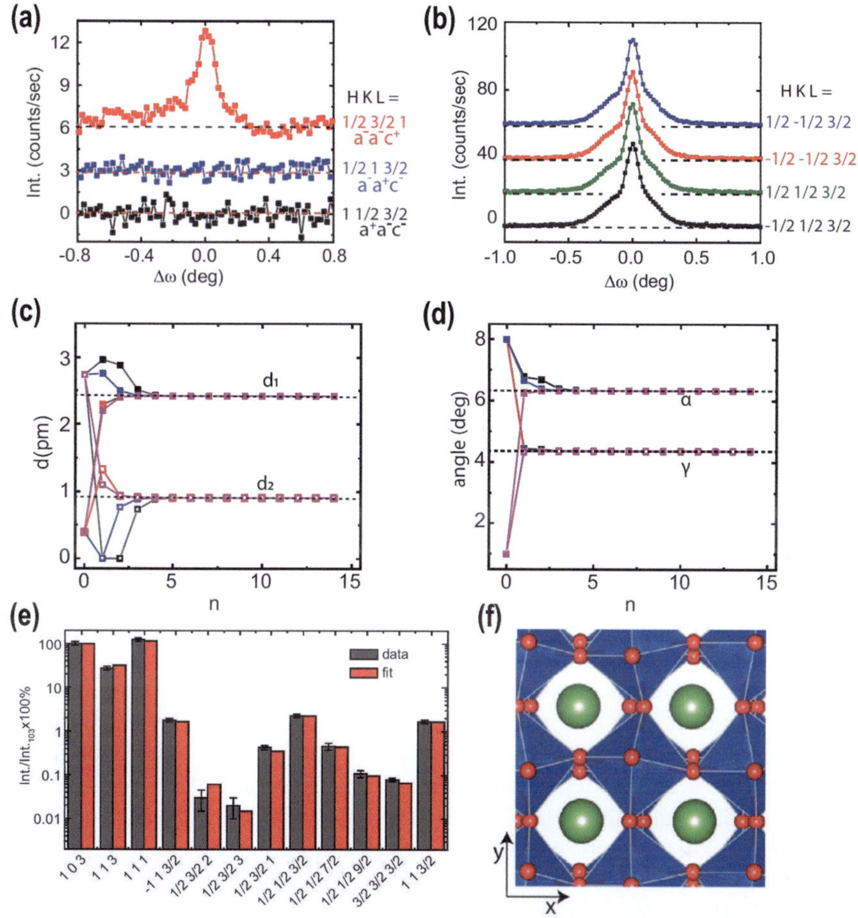

Fig. 5 Measurement and analysis of half-order peaks from a La($Cr_{0.2}Mn_{0.2}Fe_{0.2}Co_{0.2}Ni_{0.2}$)$O_3$/ $SrTiO_3$ heterostructure illustrating the steps needed to refine the atomic positions. **a** Survey scans are used to establish that the film has an $a^-a^-c^+$ pattern. **b** Measurements of the $\frac{\pm 1}{2}\ \frac{\pm 1}{2}\ \frac{3}{2}$ peaks confirm an equal population of rotational domains ($D_j = 0.25$ for all j). **c, d** The values of the d_1, d_2, α and γ displacement parameters as a function of the number of iterations of the fitting routine. **e** Comparison of the measured and simulated peak intensities after completion of the fitting procedure; note the excellent agreement even with the peaks varying in intensity by approximately five orders magnitude. **f** A schematic of the refined crystal structure of the film viewed down the [100] pseudocubic direction. All panels reproduced from Ref. [47]

substitution with a smaller *A*-site cation, such as Pr or Nd, the octahedral rotations increase in magnitude, the pattern changes to $a^-a^-c^+$, and the compound exhibits a first-order metal–insulator transition [5, 69]. The temperature of this phase transition directly correlates with the Ni–O–Ni bond angle, with decreasing θ values increasing the stability of the insulating state. Over the last decade, a significant amount of research was focused on understanding the coupling of structural and electronic properties in strained nickelate heterostructures [70], including the refinement of half-order reflections to quantify octahedral rotations in these systems.

Figure 6 provides a summary of how the α and γ rotations in $LaNiO_3$ films are impacted by strain, thickness, and interface formation in superlattices. In all these examples, the films were grown on substrates with equal in-plane lattice parameters along the *a*- and *b*-axes, and thus α is equal to β. In response to compressive strain, the γ rotation increases to accommodate the reduction in in-plane lattice parameter. The opposite response is found under tensile strain, where γ increases to straighten out the in-plane bonds to accommodate the larger in-plane lattice parameters. The α angles exhibit opposite responses under biaxial strain, as shown in Fig. 6a, d. Results from transmission electron diffraction analysis of $LaNiO_3$ on LSAT substrates are also included in Fig. 6a, revealing rotation angles in good agreement with synchrotron-based refinements. Analysis of half-order X-ray diffraction peaks has revealed similar strain-dependent trends in epitaxial $LaAlO_3$ [71], $Pr_{0.67}Sr_{0.33}MnO_3$ [72], and $La_{0.5}Sr_{0.5}CoO_3$ films [73].

Half-order diffraction peaks were also analyzed in $LaNiO_3$ films grown on $SrTiO_3$ and $LaAlO_3$ substrates to understand how octahedral rotations evolve as a function of thickness by Fowlie et al. [68] As shown in Fig. 6b, the average rotational magnitudes are slightly suppressed in ultrathin films grown on $LaAlO_3$ potentially due to the increased role of the surface as the film thickness is decreased. However, the rotation angles are essentially constant for $LaNiO_3/LaAlO_3$ as the thickness is increased beyond 6-unit cells. The minimal changes in $LaNiO_3/LaAlO_3$ heterostructures are likely due to the structural similarities between the two materials as both have the $a^-a^-a^-$ pattern in bulk with similar α values (5.2° for $LaNiO_3$ and 5.6° for $LaAlO_3$). In contrast, the symmetry mismatch between $LaNiO_3$ and $SrTiO_3$ leads to more significant thickness effects, where in 5-unit cell films the γ angle is larger than α, while for films thicker than 10-unit cells, α is much larger than γ as expected for a film under tensile strain. These results highlight the complex ways in which the symmetry mismatch can be accommodated at perovskite heterointerfaces, with potential implications for electronic and magnetic properties. For instance, the abruptness and presence of thermal hysteresis across a metal–insulator transition in $NdNiO_3$ films was correlated to increases in bond angles as the film thickness was increased [74].

Similar to how rotation angles are modified at the film/substrate interface, the growth of superlattices enables the formation of interfaces but with greater control over the constituent materials and the interfacial density. In short-period superlattices, where the distance between interfaces is greater than the structural coupling length scale, crystals with relatively uniform rotations can be engineered by tuning the layer thicknesses, enabling ultrathin layers with bond angles deviating dramatically from their bulk forms. For example, $(LaNiO_3)_n/(SrMnO_3)_2$ superlattices force interfacial connectivity between $LaNiO_3$ and cubic perovskite $SrMnO_3$ ($a^0a^0a^0$) [76]. Analysis of half-order diffraction peaks revealed that the rotations in a $(LaNiO_3)_1/(SrMnO_3)_2$ superlattice are minimal with average *B*–O–*B* bond angles on the order of 175°, in stark contrast to bulk $LaNiO_3$ where $\theta = 165.2°$. As more $LaNiO_3$ layers are added to the superlattice period, the α and γ angles are increased leading to more bulk-like

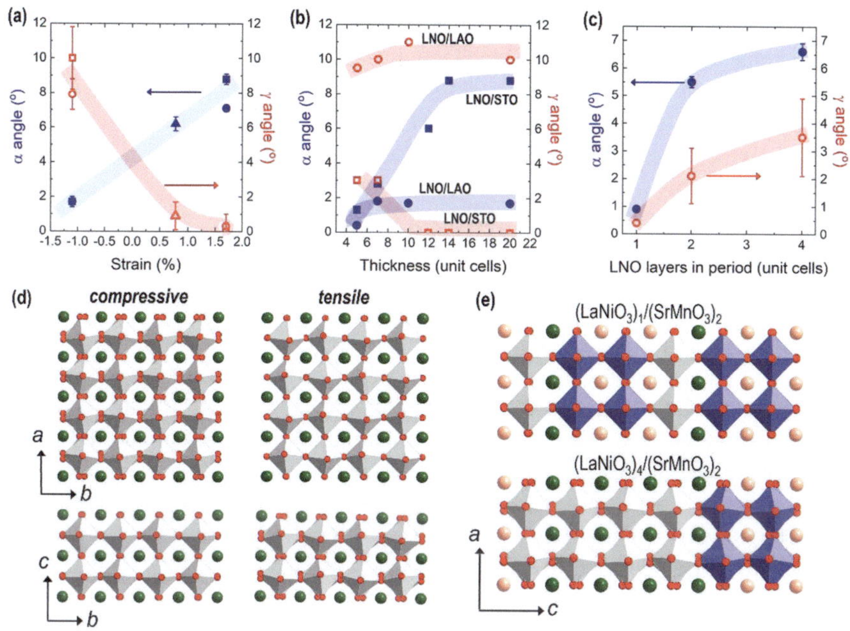

Fig. 6 Examples from $LaNiO_3$-based heterostructures of how strain, thickness, and interface formation alter octahedral rotations. **a** The α and γ rotation angles measured in $LaNiO_3$ films under different strain states. The circles and squares were measured using synchrotron diffraction; the triangles were obtained from transmission electron diffraction analysis. The data are reproduced from Refs. [16] (circles) [68], (squares), and [75] (triangles). **b** The evolution of rotations as a function of film thickness; data reproduced from Ref. [68]. **c** Rotations measured in $(LaNiO_3)_n/(SrMnO_3)_2$ superlattices; data reproduced from Ref. [76]. For **a**–**c** blue symbols indicate $\alpha = \beta$ angles and red symbols indicate γ angles. **d** Crystal structures of $LaNiO_3$ under compressive strain (on $LaAlO_3$) and tensile strain (on $SrTiO_3$) illustrate how the in-plane and out-of-plane bond angles are modified by strain. **e** Crystal structures of $(LaNiO_3)_n/(SrMnO_3)_2$ superlattices reveal the increase in rotations with increasing number of consecutive $LaNiO_3$ layers. Crystal schematics adapted from Refs. [16, 76]

bond angles, as shown in Fig. 6c, e. However, compared to strain tuning of rotations, the use of interfacial coupling either in superlattices or from substrate-effects in ultra-thin films leads to a much wider range of accessible θ angles. Octahedral rotations have also been studied in $(LaNiO_3)/(LaAlO_3)$ superlattices using a combination of synchrotron diffraction and high resolution transmission electron microscopy [77]. Another example of how octahedral rotations can be tuned in multilayers comes from analysis of $(LaMnO_3)_n/(SrTiO_3)_n$ superlattices, where Zhai and coauthors determined the rotational angles for numerous structures of differing periodicity and linked the saturation magnetization of $LaMnO_3$ to its γ angle [78, 79].

The ability to tune rotations through strain and superlattice formation has enabled new insights into the relationship between the octahedral geometry and ionic conductivity in $La_{0.9}Sr_{0.1}Ga_{0.95}Mg_{0.05}O_{3-\delta}$. Gao et al. employed a combination of synchrotron scattering, impedance spectroscopy, and density functional theory calculations to identify that ionic conductivity is increased through both increased octahedral rotations and unit cell volume [80]. However, the sole application of biaxial strain was found to increase or decrease average rotations while inducing the opposite trend in cell volume. To overcome this limitation, the authors grew $(La_{0.9}Sr_{0.1}Ga_{0.95}Mg_{0.05}O_{3-\delta})/(DyScO_3)$ superlattices under tensile strain, whereby structural coupling with $DyScO_3$ increased the rotations in the gallate layer while the unit cell volume was increased due to the strain.

Here it is worth recalling that the rotation angles obtained from the measurement and analysis of half-order peaks are average values as basic refinement of diffraction peaks does not allow for resolution of depth dependence or layer-by-layer evolution of rotations within a single film. As the superlattice period is increased and rotations relax from interfacially-imposed geometric constraints, the α and γ values will vary spatially through the thickness of the heterostructure. This additional periodicity arising from modulations of the rotational magnitudes, on top of the basic unit cell doubling nature of the rotations, leads to satellite peaks off of the half-order reflections [24]. The position of the satellite peaks along L relative to the half-order peaks (δ in $Å^{-1}$) is inversely related to the modulation period (Δ), through $\Delta = 2\pi/\delta$, as is standard for superstructure reflections. In principle, the amplitude of the rotational modulations could be determined through quantitative analysis of the intensity ratios between the half-order and satellite peaks, but I'm unaware of any published work yet carrying out this approach.

In situ response of octahedral rotations to external stimuli. Beyond measuring static structures, recent studies have demonstrated how synchrotron diffraction can be used to measure dynamic responses to external stimuli in perovskite films. The application of electrolytic gating has been applied widely within the oxide heterostructure community as a means to alter functional properties of perovskite films through either electrostatic carrier modulation or electrochemical vacancy insertion/removal [81]. While octahedral rotations are largely non-responsive to electric fields, in cases where a gate bias alters the oxygen content within a film, the octahedral network can respond through changes to the average rotational magnitudes.

Liu and coauthors demonstrated this strategy for electric field control of octahedral rotations in $WO_{3-\delta}$ thin films gated with an ionic liquid [82]. $WO_{3-\delta}$ crystallizes in a corner-sharing network of octahedra, equivalent to a perovskite without A-site cations. The rotation pattern for bulk WO_3 is $a^-b^+c^-$. In the case of $WO_{3-\delta}$, the presence of a positive gate bias removes oxygen from the film, leading to a decrease in resistivity, while a negative bias has minimal effects. The structural consequences of film reduction were directly measured by tracking the intensity of numerous half-order diffraction peaks as a function of time while the gate bias was applied and removed, as presented in Fig. 7. The authors found that the in-plane b^+ rotation is not affected as the $\frac{1}{2}\,0\,\frac{3}{2}$ peak intensity is constant. In contrast, the $\frac{1}{2}\,\frac{3}{2}\,\frac{3}{2}$ intensity decreased by 30% indicating a significant suppression of γ rotation along the c^- axis. The oxygen loss also resulted in displacements of the W cations, which were tracked through measurement of $\frac{1}{2}\,0\,2$ and $\frac{1}{2}\,\frac{3}{2}\,2$ peaks. To confirm that these peaks arise from

W displacements, and that the $\frac{1}{2}\,\frac{3}{2}\,\frac{3}{2}$ and $\frac{1}{2}\,0\,\frac{3}{2}$ peaks do not, energy scans at the W L_3-edge were performed revealing a resonance at the former set of peaks and no energy dependence at the latter sets of peaks. The authors used spectroscopic measurements to quantify the oxygen content as a function of bias, which was then used to calculate the individual contributions of the suppressed c^- rotation and the decreased oxygen occupancy to the measured $\frac{1}{2}\,\frac{3}{2}\,\frac{3}{2}$ intensity. Complementary density functional theory calculations found a similar relationship between the oxygen content in $WO_{3-\delta}$ and the magnitude of its octahedral rotations.

The structural impacts of optical excitation were investigated in $LaVO_3$ films using time-resolved pump-probe measurements of half-order reflections [83]. $LaVO_3$ is an orthorhombic perovskite with both octahedral rotations ($a^-a^-c^+$) and *A*-site displacements. Upon excitation, the *c*-axis parameter increased by 0.1% followed by a gradual decrease back to the unexcited lattice volume over the course of tens of ns. The impact on the octahedral network was quantified by measuring the intensities of the $\frac{1}{2}\,\frac{3}{2}\,\frac{3}{2}$ and $\frac{1}{2}\,\frac{3}{2}\,\frac{3}{2}$ peaks, while the *A*-site displacements were tracked via the $0\,\frac{1}{2}\,1$ and $\frac{3}{2}\,\frac{3}{2}\,\frac{3}{2}$ peaks. Through analysis of the static structure and then by tracking time-resolved changes to these specific reflections, the authors were able to quantify the excitation-induced modifications to the atomic positions revealing a 0.3° decrease in both α and γ and decreases of 0.14 and 0.18 pm to d_1 and d_2, respectively, within 1 ns of optical pumping. These reductions in displacements are consistent with thermal effects pushing the sample closer to the cubic, undistorted perovskite structure.

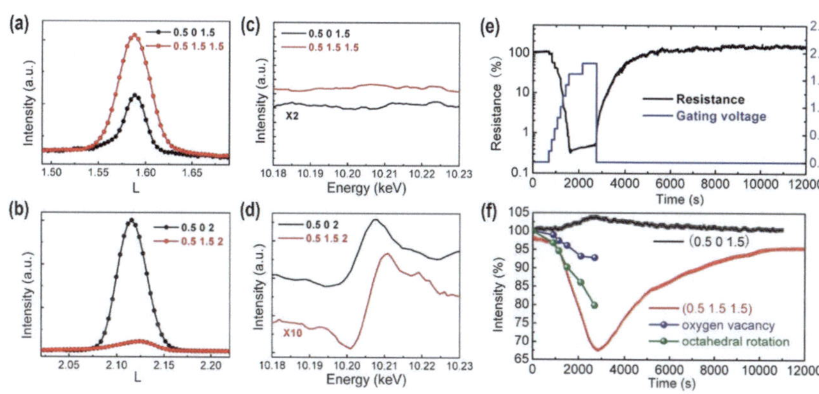

Fig. 7 Measurement of half-order peaks arising from **a** octahedral rotations and **b** *B*-site displacements in $WO_{3-\delta}$ films. **c, d** Energy scans across the W L_3-edge confirm that the peaks in **a** do not contain contributions from W displacements, while the peaks in **b** arise from W displacements. **e** The application of a bias across an ionic liquid gate decreases the resistivity of the film by increasing the oxygen vacancy concentration. **f** The structural impact can be tracked through measurements of half-order peaks, which reveal a significant decrease in the γ rotation with application of the gate bias. The decrease in oxygen occupancy also contributes to the reduction in half-order intensity. Figures reproduced from Ref. [82]

4 Summary—Future Opportunities

Looking to the future, there are numerous opportunities to exploit half-order diffraction peaks to extract new information from complex oxide heterostructures. Analysis and simulations of full crystal truncation rods through half-order reflections, for instance via application of the coherent Bragg rod analysis (COBRA) method [84], can provide depth-dependent information yielding layer-by-layer rotation angles as opposed to macroscopic averages. Initial work in this area has been performed by Kan and coauthors, who measured and simulated truncation rods with h and k equal to half-order indices to determine the tilt patterns and average rotational magnitudes of $SrRuO_3$ films [48]. Yuan and coauthors were able to obtain layer-resolved oxygen positions in thin epitaxial $CaTiO_3$ films using COBRA measured through integer h and k indices [34].

Resonant diffraction, in which the photon energy is scanned across or tuned to an electronic transition to alter the form factor f, provides a means of distinguishing or highlighting elemental contributions to Bragg reflections [85, 86]. This approach was used in the previously mentioned study of $WO_{3-\delta}$ to distinguish between W and O displacements [82]. Resonant scattering of half-order peaks at the K- or L-edge transition of A-site cations may be useful for refining displacements particular to specific cations in $A_{1-x}A'_xBO_3$ films or $(ABO_3)/(A'BO_3)$ superlattices. Quantitative measurements of A-site displacements in short-period $(ABO_3)/(A'BO_3)$ superlattices would be particularly useful for verifying predictions of hybrid improper ferroelectric states in these structures [22, 87]. Techniques that take advantage of the beam coherence may also find use in studying aspects of domains formed from octahedral rotations. For instance, the dynamics of domains fluctuations near structural phase transitions could be investigated using X-ray photon correlation spectroscopy [88, 89], while coherent diffractive imaging [90, 91] may yield real space insights into orientational domains or how octahedral rotations are modified near interfaces.

Finally, the approaches outlined within this chapter can be applied to epitaxial heteroanionic perovskites or antiperovskites. Beyond ABO_3 oxides, there are numerous oxynitrides and oxyfluorides that can be stabilized as epitaxial films either in an as-deposited state or through post-growth topochemical transformations [92]. In some cases, epitaxial strain has been demonstrated to promote anion-site ordering, for instance with either nitrogen or fluorine occupying the apical or equatorial sites in oxynitrides or oxyfluorides [93, 94]. Synchrotron diffraction studies of half-order peaks would shed light on the bonding geometries that result in these structures. In A_3BO antiperovskites, such as Sr_3SnO, the crystal structure contains corner-connected OA_6 octahedral units. As these materials are of growing interest for hosting topological bands, superconductivity and promising thermoelectric properties [95, 96], half-order diffraction studies may yield new insights regarding how to tune topological band features through strain or interface-induced modifications to the octahedral network in antiperovskites heterostructures [97, 98].

Acknowledgements I am deeply grateful to Matthew Brahlek, Dillon Fong, Jennifer Fowlie, John Freeland, Jenia Karapetrova, Jong-Woo Kim, and Phil Ryan for insightful discussions related to synchrotron scattering measurements of perovskite oxide heterostructures. I also thank the National Science Foundation (grant #DMR-1151649) and the Army Research Office (grants #W911NF-12-1-0132; #W911NF-15-1-0133) for supporting my work on octahedral rotations in oxide heterostructures.

Questions and Problems (Measuring Octahedral Rotations in Perovskite Oxide Heterostructures)

Q 6.1.

i. Consider an ideal cubic perovskite with the lattice parameter *a*. Show that the *B*–O bond length *d* after distorting the *B*–O–*B* bond angle along one direction to an angle θ (Fig. 1b) is given by:

$$d = \frac{a}{2 \sin \frac{\theta}{2}}$$

ii. Show that the distance between the original O position and new O position, *x*, is given by:

$$x = \frac{2}{a} \tan \alpha$$

where α is an octahedral rotation around an axis perpendicular to the *B*–O bond.

Q 6.2.

i. A scientist is researching a (001)-oriented perovskite film that is known to have the $a^-a^-c^+$ (*Pbnm*) rotation pattern but is not sure whether the in-phase rotations are along [001] (the out-of-plane direction), [010], or [100] (the in-plane directions). The film is on a cubic substrate. Measurements reveal the presence of the $\frac{3}{2}\,\frac{1}{2}\,\frac{1}{2}$ and $\frac{1}{2}\,\frac{1}{2}\,\frac{3}{2}$ reflections. The scientist checks Table 1 and concludes that the film must have out-of-phase rotations along [010]. Is the presence of these two peaks sufficient to support this claim? If not, what else needs to be measured?
ii. The scientist grows the same film on two different substrates: $NdGaO_3$ and $LaAlO_3$. How might the rotation pattern of the film differ between these two samples?

Q 6.3. A student submits a paper in which they have refined the oxygen positions of a *Pbnm* $a^-a^-c^+$ perovskite oxide film by comparing measured intensity of half-order to simulations based on oxygen displacements. Reviewers argue that more measurements or analysis need to be performed in order to obtain the full crystal structure, even though the oxygen positions were refined. What is the student missing?

Q 6.4. Resonant scattering, in which the photon energy is tuned to the *L*-, *K*-, or *M*-edge of the cation, has been used to assist in the refinement of *A*-site positions. Can resonant scattering be used to assist in the refinement of oxygen positions?

References

1. A.M. Glazer, Classification of tilted octahedra in perovskites. Acta Crystallogr. B **28**, 3384 (1972)
2. P.M. Woodward, Octahedral tilting in perovskites. I. Geometrical considerations. Acta Crystallogr. B **53**, 32 (1997)
3. P.M. Woodward, Octahedral tilting in perovskites. II. Structure stabilizing forces. Acta Crystallogr. B **53**, 44 (1997)
4. C.J. Howard, H.T. Stokes, Group-theoretical analysis of octahedral tilting in perovskites. Acta Crystallogr. B **54**, 782 (1998)
5. J.B. Torrance, P. Lacorre, A.I. Nazzal, E.J. Ansaldo, C. Niedermayer, Systematic study of insulator-metal transitions in perovskites $RNiO_3$ (R = Pr, Nd, Sm, Eu) due to closing of charge-transfer gap. Phys. Rev. B **45**, 8209 (1992)
6. T. Takeda, R. Kanno, Y. Kawamoto, M. Takano, S. Kawasaki, T. Kamiyama, F. Izumi, Metal-semiconductor transition, charge disproportionation, and low-temperature structure of $Ca_{1-x}Sr_xFeO_3$ synthesized under high-oxygen pressure. Solid State Sci. **2**, 673 (2000)
7. Y. Tomioka, Y. Tokura, Global phase diagram of perovskite manganites in the plane of quenched disorder versus one-electron bandwidth. Phys. Rev. B **70**, 014432 (2004)
8. P.G. Radaelli, G. Iannone, M. Marezio, H.Y. Hwang, S.-W. Cheong, J.D. Jorgensen, D.N. Argyrious, Structural effects on the magnetic and transport properties of perovskite $A_{1-x}A_x{}'MnO_3$ ($x = 0.25, 0.30$). Phys. Rev. B **56**, 8265 (1997)
9. J.S. Zhou, J.B. Goodenough, Unusual evolution of the magnetic interactions versus structural distortions in $RMnO_3$ perovskites. Phys. Rev. Lett. **96**, 247202 (2006)
10. H.W. Eng, P.W. Barnes, B.M. Auer, P.M. Woodward, Investigations of the electronic structure of d^0 transition metal oxides belonging to the perovskite family. J. Solid State Chem. **175**, 94 (2003)
11. S. Lee, W.H. Woodford, C.A. Randall, Crystal and defect chemistry influences on band gap trends in alkaline earth perovskites. Appl. Phys. Lett. **92**, 201909 (2008)
12. J.M. Rondinelli, S.J. May, J.W. Freeland, Control of octahedral connectivity in perovskite oxide heterostructures: an emerging route to multifunctional materials discovery. MRS Bull. **37**, 261 (2012)
13. J.M. Rondinelli, N.A. Spaldin, Structure and properties of functional oxide thin films: insights from electronic-structure calculations. Adv. Mater. **23**, 3363 (2011)
14. D.G. Schlom, L.-Q. Chen, C.J. Fennie, V. Gopalan, D.A. Muller, X. Pan, R. Ramesh, R. Uecker, Elastic strain engineering of ferroic oxides. MRS Bull. **39**, 118 (2014)
15. A.T. Zayak, X. Huang, J.B. Neaton, K.M. Rabe, Structural, electronic, and magnetic properties of $SrRuO_3$ under epitaxial strain. Phys. Rev. B **74**, 094104 (2006)
16. S.J. May, J.W. Kim, J.M. Rondinelli, E. Karapetrova, N.A. Spaldin, A. Bhattacharya, P.J. Ryan, Quantifying octahedral rotations in strained perovskite oxide films. Phys. Rev. B **82**, 014110 (2010)

17. C.L. Jia, S.B. Mi, M. Faley, U. Poppe, J. Schubert, K. Urban, Oxygen octahedron reconstruction in the $SrTiO_3/LaAlO_3$ heterointerfaces investigated using aberration-corrected ultrahigh-resolution transmission electron microscopy. Phys. Rev. B **79**, 081405R (2009)
18. A.Y. Borisevich et al., Suppression of octahedral tilts and associated changes in electronic properties at epitaxial oxide heterostructure interfaces. Phys. Rev. Lett. **105**, 087204 (2010)
19. J.M. Rondinelli, N.A. Spaldin, Substrate coherency driven octahedral rotations in perovskite oxide films. Phys. Rev. B **82**, 113402 (2010)
20. R. Aso, D. Kan, Y. Shimakawa, H. Kurata, Octahedral tilt propagation controlled by A-site cation size at perovskite oxide heterointerfaces. Cryst. Growth Des. **14**, 2128 (2014)
21. E. Bousquet, M. Dawber, N. Stucki, C. Lichtensteiger, P. Hermet, S. Gariglio, J.M. Triscone, P. Ghosez, Improper ferroelectricity in perovskite oxide artificial superlattices. Nature **452**, 732 (2008)
22. J.M. Rondinelli, C.J. Fennie, Octahedral rotation-induced ferroelectricity in cation ordered perovskites. Adv. Mater. **24**, 1961 (2012)
23. S. Ghosh, H. Das, C.J. Fennie, Linear magnetoelectricity at room temperature in perovskite superlattices by design. Phys. Rev. B **92**, 184112 (2015)
24. E.J. Moon, R. Colby, Q. Wang, E. Karapetrova, C.M. Schleputz, M.R. Fitzsimmons, S.J. May, Spatial control of functional properties via octahedral modulations in complex oxide superlattices. Nat. Commun. **5**, 5710 (2014)
25. S. Catalano et al., Tailoring the electronic transitions of $NdNiO_3$ films through $(111)_{pc}$ oriented interfaces. APL Mater. **3**, 062506 (2015)
26. B. Chen et al., Spatially controlled octahedral rotations and metal-insulator transitions in nickelate superlattices. Nano Lett. **21**, 1295 (2021)
27. D. Kan, R. Aso, R. Sato, M. Haruta, H. Kurata, Y. Shimakawa, Tuning magnetic anisotropy by interfacially engineering the oxygen coordination environment in a transition metal oxide. Nat. Mater. **15**, 432 (2016)
28. Z. Liao et al., Controlled lateral anisotropy in correlated manganite heterostructures by interface-engineered oxygen octahedral coupling. Nat. Mater. **15**, 425 (2016)
29. D. Puggioni, J.M. Rondinelli, Designing a robustly metallic noncenstrosymmetric ruthenate oxide with large thermopower anisotropy. Nat. Commun. **5**, 3432 (2014)
30. T.H. Kim et al., Polar metals by geometric design. Nature **533**, 68 (2016)
31. E.J. Moon, Q. He, S. Ghosh, B.J. Kirby, S.T. Pantelides, A.Y. Borisevich, S.J. May, Structural "δ doping" to control local magnetization in isovalent oxide heterostructures. Phys. Rev. Lett. **119**, 197204 (2017)
32. T.T. Fister et al., Octahedral rotations in strained $LaAlO_3/SrTiO_3$ (001) heterostructures. APL Mater. **2**, 021102 (2014)
33. D.P. Kumah, A.S. Disa, J.H. Ngai, H.H. Chen, A. Malashevich, J.W. Reiner, S. Ismail-Beigi, F.J. Walker, C.H. Ahn, Tuning the structure of nickelates to achieve two-dimensional electron conduction. Adv. Mater. **26**, 1935 (2014)
34. Y. Yuan, Y. Lu, G. Stone, K. Wang, C.M. Brooks, D.G. Schlom, S.B. Sinnott, H. Zhou, V. Gopalan, Three-dimensional atomic scale electron density reconstruction of octahedral tilt epitaxy in functional perovskites. Nat. Commun. **9**, 5220 (2018)
35. A.M. Glazer, Simple ways of determining perovskite structures. Acta Crystallogr. A **31**, 756 (1975)
36. J.W. Seo, E.E. Fullerton, F. Nolting, A. Scholl, J. Fompeyrine, J.P. Locquet, Antiferromagnetic $LaFeO_3$ thin films and their effect on exchange bias. J. Phys. Condens. Matter **20**, 264014 (2008)
37. D.L. Proffit, H.W. Jang, S. Lee, C.T. Nelson, X.Q. Pan, M.S. Rzchowski, C.B. Eom, Influence of symmetry mismatch on heteroepitaxial growth of perovskite thin films. Appl. Phys. Lett. **93**, 111912 (2008)
38. H. Rotella, U. Lüders, P.E. Janolin, V.H. Dao, D. Chateigner, R. Feyerherm, E. Dudzik, W. Prellier, Octahedral tilting in strained $LaVO_3$ thin films. Phys. Rev. B **85**, 184101 (2012)
39. A.K. Choquette, C.R. Smith, R.J. Sichel-Tissot, E.J. Moon, M.D. Scafetta, E. Di Gennaro, F. Miletto Granozio, E. Karapetrova, S.J. May, Octahedral rotation patterns in strained $EuFeO_3$

and other Pbnm perovskite films: implications for hybrid improper ferroelectricity. Phys. Rev. B **94**, 024105 (2016)
40. R. Gao et al., Interfacial octahedral rotation mismatch control of the symmetry and properties of $SrRuO_3$. ACS Appl. Mater. Interfaces **8**, 14871 (2016)
41. C. Lau, Y. Kim, S. Albright, K. Char, C.H. Ahn, F.J. Walker, Structural characterization of the $LaInO_3/BaSnO_3$ interface via synchrotron scattering. APL Mater. **7**, 031108 (2019)
42. T.C. van Thiel, J. Fowlie, C. Autieri, N. Manca, M. Šiškins, D. Afanasiev, S. Gariglio, A.D. Caviglia, Coupling lattice instabilities across the interface in ultrathin oxide heterostructures. ACS Mater. Lett. **2**, 389 (2020)
43. H. Meley, Karandeep, L. Oberson, J. de Bruijckere, D.T.L. Alexander, J.M. Triscone, P. Ghosez, S. Gariglio, Structural analysis of $LaVO_3$ thin films under epitaxial strain. APL Mater. **6**, 046102 (2018)
44. M.D. Biegalski, L. Qiao, Y. Gu, A. Mehta, Q. He, Y. Takamura, A. Borisevich, L.-Q. Chen, Impact of symmetry on the ferroelectric properties of $CaTiO_3$ thin films. Appl. Phys. Lett. **106**, 162904 (2015)
45. D. Kan, Y. Wakabayashi, H. Tajiri, Y. Shimakawa, Interfacially engineered oxygen octahedral rotations and their impact on strain relief in coherently grown $SrRuO_3$ films. Phys. Rev. B **94**, 024112 (2016)
46. W.T. Jin, H. Gretarsson, S.Y. Jang, C.-Y. Kim, T.W. Noh, Y.-J. Kim, Hidden structural transition in epitaxial $Ca_{0.5}Sr_{0.5}IrO_3/SrTiO_3$ thin film. Phys. Rev. Mater. **3**, 103604 (2019)
47. M. Brahlek, A.R. Mazza, K.C. Pitike, E. Skoropata, J. Lapano, G. Eres, V.R. Cooper, T.Z. Ward, Unexpected crystalline homogeneity from the disordered bond network in $La(Cr_{0.2}Mn_{0.2}Fe_{0.2}Co_{0.2}Ni_{0.2})O_3$ films. Phys. Rev. Mater. **4**, 054407 (2020)
48. D. Kan, M. Anada, Y. Wakabayashi, H. Tajiri, Y. Shimakawa, Oxygen octahedral distortions in compressively strained $SrRuO_3$ epitaxial thin films. J. Appl. Phys. **123**, 235303 (2018)
49. P.F. Chen et al., Correlated cation lattice symmetry and oxygen octahedral rotation in perovskite oxide heterostructures. J. Appl. Phys. **129**, 025303 (2021)
50. F. He, B.O. Wells, Z.G. Ban, S.P. Alpay, S. Grenier, S.M. Shapiro, W. Si, A. Clark, X.X. Xi, Structural phase transition in epitaxial perovskite films. Phys. Rev. B **70**, 235405 (2004)
51. F. He, B.O. Wells, S.M. Shapiro, Strain phase diagram and domain orientation in $SrTiO_3$ thin films. Phys. Rev. Lett. **94**, 176101 (2005)
52. R.L. Johnson-Wilke, D.S. Tinberg, C.B. Yeager, Y. Han, I.M. Reaney, I. Levin, D.D. Fong, T.T. Fister, S. Trolier-McKinstry, Tilt transitions in compressively strained $AgTa_{0.5}Nb_{0.5}O_3$ thin films. Phys. Rev. B **84**, 134114 (2011)
53. R.L. Johnson-Wilke, R.H.T. Wilke, C.B. Yeager, D.S. Tinberg, I.M. Reaney, I. Levin, D.D. Fong, S. Trolier-McKinstry, Phase transitions and octahedral rotations in epitaxial $Ag(Ta_xNb_{1-x})O_3$ thin films under tensile strain. J. Appl. Phys. **117**, 085309 (2015)
54. M. Ziese, I. Vrejoiu, D. Hesse, Structural symmetry and magnetocrystalline anisotropy of $SrRuO_3$ films on $SrTiO_3$. Phys. Rev. B **81**, 184418 (2010)
55. D. Kan, R. Aso, H. Kurata, Y. Shimakawa, Thickness-dependent structure-property relationships in strained (110) $SrRuO_3$ thin films. Adv. Funct. Mater. **23**, 1129 (2013)
56. G. Koster, L. Klein, W. Siemons, G. Rijnders, J.S. Dodge, C.-B. Eom, D.H.A. Blank, M.R. Beasley, Structure, physical properties, and applications of $SrRuO_3$ thin films. Rev. Mod. Phys. **84**, 253 (2012)
57. S.H. Chang, Y.J. Chang, S.Y. Jang, D.W. Jeong, C.U. Jung, Y.J. Kim, J.S. Chung, T.W. Noh, Thickness-dependent structural phase transition of strained $SrRuO_3$ ultrathin films: the role of octahedral tilt. Phys. Rev. B **84**, 104101 (2011)
58. W. Lu, P. Yang, W.D. Song, G.M. Chow, J.S. Chen, Control of oxygen octahedral rotations and physical properties in $SrRuO_3$ films. Phys. Rev. B **88**, 214115 (2013)
59. A. Herklotz, A.T. Wong, T. Meyer, M.D. Biegalski, H.N. Lee, T.Z. Ward, Controlling octahedral rotations in a perovskite via strain doping. Sci. Rep. **6**, 26491 (2016)
60. Y. Gu et al., Interfacial oxygen-octahedral-tilting-driven electrically tunable topological Hall effect in ultrathin $SrRuO_3$ films. J. Phys. D Appl. Phys. **52**, 404001 (2019)

61. S. Lin et al., Dimensional control of octahedral tilt in $SrRuO_3$ via infinite-layered oxides. Nano Lett. **21**, 3146 (2021)
62. A.J. Grutter, A. Vailionis, J.A. Borchers, B.J. Kirby, C.L. Flint, C. He, E. Arenholz, Y. Suzuki, Interfacial symmetry control of emergent ferromagnetism at the nanoscale. Nano Lett. **16**, 5647 (2016)
63. D. Meyers et al., Magnetism in iridate heterostructures leveraged by structural distortions. Sci. Rep. **9**, 4263 (2019)
64. A. de Andrés, J. Rubio, G. Castro, S. Taboada, J.L. Martínez, J.M. Colino, Structural and magnetic properties of ultrathin epitaxial $La_{0.7}Ca_{0.3}MnO_3$ manganite films: strain versus finite size effects. Appl. Phys. Lett. **83**, 713 (2003)
65. E. Hovestreydt, On the atomic scattering factor for O^{2-}. Acta Crystallogr. A **39**, 268 (1983)
66. P.J. Brown, A.G. Fox, E.N. Maslen, M.A. O'Keefe, B.T.M. Willis, *International Tables for Crystallography*, vol. C (2006)
67. M. Brahlek, A.K. Choquette, C.R. Smith, R. Engel-Herbert, S.J. May, Structural refinement of Pbnm-type perovskite films from analysis of half-order diffraction peaks. J. Appl. Phys. **121**, 045303 (2017)
68. J. Fowlie, C. Lichtensteiger, M. Gibert, H. Meley, P. Willmott, J.-M. Triscone, Thickness-dependent perovskite octahedral distortions at heterointerfaces. Nano Lett. **19**, 4188 (2019)
69. M.L. Medarde, Structural, magnetic and electronic properties of perovskites (R = rare earth). J. Phys. Condens. Matter **9**, 1679 (1997)
70. S. Catalano, M. Gibert, J. Fowlie, J. Íñiguez, J.M. Triscone, J. Kreisel, Rare-earth nickelates $RNiO_3$: thin films and heterostructures. Rep. Prog. Phys. **81**, 046501 (2018)
71. R.L. Johnson-Wilke et al., Quantification of octahedral rotations in strained $LaAlO_3$ films via synchrotron X-ray diffraction. Phys. Rev. B **88**, 174101 (2013)
72. B. Zhang, C.-J. Sun, P. Yang, W. Lu, B.L. Fisher, T. Venkatesan, S.M. Heald, J.-S. Chen, G.M. Chow, Strain modulated anisotropic electronic charge transfer in perovskite $Pr_{0.67}Sr_{0.33}MnO_3$ thin films. Phys. Rev. B **89**, 195140 (2014)
73. Y. Heo, D. Kan, M. Anada, Y. Wakabayashi, H. Tajiri, Y. Shimakawa, Correlations between oxygen octahedral distortions and magnetic and transport properties in strained $La_{0.5}Sr_{0.5}CoO_3$ thin films. Phys. Rev. B **99**, 174420 (2019)
74. Y. Dong, Z. Ma, Z. Luo, H. Zhou, D.D. Fong, W. Wu, C. Gao, Interfacial octahedral manipulation imparts hysteresis-free metal to insulator transition in ultrathin nickelate heterostructure. Adv. Mater. Interfaces **6**, 1900644 (2019)
75. J. Hwang, J.Y. Zhang, J. Son, S. Stemmer, Nanoscale quantification of octahedral tilts in perovskite films. Appl. Phys. Lett. **100**, 191909 (2012)
76. S.J. May, C.R. Smith, J.W. Kim, E. Karapetrova, A. Bhattacharya, P.J. Ryan, Control of octahedral rotations in $(LaNiO_3)_n/(SrMnO_3)_m$ superlattices. Phys. Rev. B **83**, 153411 (2011)
77. M.K. Kinyanjui et al., Lattice distortions and octahedral rotations in epitaxially strained $LaNiO_3/LaAlO_3$ superlattices. Appl. Phys. Lett. **104**, 221909 (2014)
78. X. Zhai et al., Correlating interfacial octahedral rotations with magnetism in $(LaMnO_{3+\delta})_N/(SrTiO_3)_N$ superlattices. Nat. Commun. **5**, 4283 (2014)
79. H. Cao, Y. Liu, G. Liang, A. Zhao, X. Zhai, Unraveling interfacial strain and interfacial lattice reconstruction mechanism of ultrathin $LaMnO_{3+\delta}$ layers in $LaMnO_{3+\delta}/SrTiO_3$ superlattices. J. Appl. Phys. **122**, 085309 (2017)
80. R. Gao et al., Designing optimal perovskite structure for high ionic conduction. Adv. Mater. **32**, 1905178 (2020)
81. C. Leighton, Electrolyte-based ionic control of functional oxides. Nat. Mater. **18**, 13 (2019)
82. H. Liu, Y. Dong, D. Xu, E. Karapetrova, S. Lee, L. Stan, P. Zapol, H. Zhou, D.D. Fong, Dynamic field modulation of the octahedral framework in metal oxide heterostructures. Adv. Mater. **30**, 1804775 (2018)
83. M. Brahlek et al., Structural dynamics of $LaVO_3$ on the nanosecond time scale. Struct. Dyn. **6**, 014502 (2019)
84. Y. Yacoby, M. Sowwan, E. Stern, J.O. Cross, D. Brewe, R. Pindak, J. Pitney, E.M. Dufresne, R. Clarke, Direct determination of epitaxial interface structure in Gd_2O_3 passivation of GaAs. Nat. Mater. **1**, 99 (2002)

85. J.-L. Hodeau, V. Favre-Nicolin, S. Bos, H. Renevier, E. Lorenzo, J.-F. Berar, Resonant diffraction. Chem. Rev. **101**, 1843 (2001)
86. Y. Lu et al., Quantitative determination of bond order and lattice distortions in nickel oxide heterostructures by resonant X-ray scattering. Phys. Rev. B **93**, 165121 (2016)
87. A.T. Mulder, N.A. Benedek, J.M. Rondinelli, C.J. Fennie, Turning ABO_3 antiferroelectrics into ferroelectrics: design rules for practical rotation-driven ferroelectricity in double perovskites and $A_3B_2O_7$ Ruddlesden-Popper compounds. Adv. Funct. Mater. **23**, 4810 (2013)
88. R. Kukreja, N. Hua, J. Ruby, A. Barbour, W. Hu, C. Mazzoli, S. Wilkins, E.E. Fullerton, O.G. Shpyrko, Orbital domain dynamics in magnetite below the Verwey transition. Phys. Rev. Lett. **121**, 177601 (2018)
89. V. Thampy et al., Static charge-density-wave order in the superconducting state of $La_{2-x}Ba_xCuO_4$. Phys. Rev. B **95**, 241111 (2017)
90. A. Yau, W. Cha, M.W. Kanan, G.B. Stephenson, A. Ulvestad, Bragg coherent diffractive imaging of single-grain defect dynamics in polycrystalline films. Science **356**, 739 (2017)
91. D. Karpov, Z. Liu, T.D.S. Rolo, R. Harder, P.V. Balachandran, D. Xue, T. Lookman, E. Fohtung, Three-dimensional imaging of vortex structure in a ferroelectric nanoparticle driven by an electric field. Nat. Commun. **8**, 280 (2017)
92. H. Kageyama, K. Hayashi, K. Maeda, J.P. Attfield, Z. Hiroi, J.M. Rondinelli, K.R. Poeppelmeier, Expanding frontiers in materials chemistry and physics with multiple anions. Nat. Commun. **9**, 772 (2018)
93. D. Oka et al., Strain engineering for anion arrangement in perovskite oxynitrides. ACS Nano **11**, 3860 (2017)
94. J. Wang et al., Strain-induced anion-site occupancy in perovskite oxyfluoride films. Chem. Mater. **33**, 1811 (2021)
95. A.W. Rost et al., Inverse-perovskites A_3BO (A = Sr, Ca, Eu/B = Pb, Sn): a platform for control of Dirac and Weyl fermions. APL Mater. **7**, 121114 (2019)
96. M. Ochi, K. Kuroki, Comparative first-principles study of antiperovskite oxides and nitrides as thermoelectric material: multiple Dirac cones, low-dimensional band dispersion, and high valley degeneracy. Phys. Rev. Appl. **12**, 034009 (2019)
97. D. Samal, H. Nakamura, H. Takagi, Molecular beam epitaxy of three-dimensional Dirac material Sr_3PbO. APL Mater. **4**, 076101 (2016)
98. Y. Ma et al., Realization of epitaxial thin films of the topological crystalline insulator Sr_3SnO. Adv. Mater. **32**, 2000809 (2020)

Synchrotron Time-Resolved X-Ray Diffraction for Piezoelectric Response of Ferroelectric Thin Films

Tomoaki Yamada, Hiroshi Funakubo, and Osami Sakata

Abstract Response of materials to the external field such as electric field, magnetic field, and mechanical force is a major concern in materials science. In-situ time-resolved X-ray diffraction (XRD) technique provides information on dynamic structural change by such external fields. In this chapter, we first introduce the basic pump and probe methods and then describe the experimental setup, data acquisition and analysis for the case of piezoelectric response of ferroelectric thin films by application of short voltage pulses at synchrotron facility.

Keywords Synchrotron · Time-resolved XRD · Ferroelectric · Piezoelectric

1 Introduction

Response of materials to the external field such as electric field, magnetic field, and mechanical force is a major concern in materials science. The strength and speed of these responses are often influenced by not only the material's composition but also its structural characteristics such as the form (ceramic, single crystal, film, or others), crystallinity, crystalline size, crystal orientation, domains, and strain. Since several different response mechanisms can interplay with each other in the materials,

T. Yamada (✉)
Department of Energy Engineering, Nagoya University, Nagoya, Japan
e-mail: yamada.tomoaki.h5@f.mail.nagoya-u.ac.jp

MDX Research Center for Element Strategy, International Research Frontiers Initiative, Institute of Science Tokyo, Yokohama, Japan

H. Funakubo
Department of Materials Science and Engineering, Institute of Science Tokyo, Yokohama, Japan
e-mail: funakubo.h.aa@m.titech.ac.jp

O. Sakata
Center for Synchrotron Radiation Research, Japan Synchrotron Radiation Research Institute, Sayo, Japan
e-mail: sakata.osami@jasri.jp

D. Sando et al. (eds.), *X-Ray Scattering Techniques for Epitaxial Oxide Thin Films*,
https://doi.org/10.1007/978-981-96-5945-6_7

the clarification of these responses on the level of crystal structure as a function of time is crucial for the development of materials.

Among various time-resolved characterization methods, X-ray diffraction (XRD) technique offers several attractive advantages. First, it is normally non-destructive. Second, it has the ability to give a range of information about the crystal structure of the sample. Third, depending on the need, one can reduce the time resolution down to a few ns or below using the appropriate facilities and setups. Fourth, it is possible to combine other measurement techniques such as electrical characterizations within the capability of the facilities used so that both the dynamic changes in the crystal structure and the property of interest can be simultaneously investigated.

One of the material groups, for which time-resolved XRD can be effectively utilized, is ferroelectrics. Various properties shown by ferroelectric materials, such as piezoelectricity, motion of domain boundaries, phase transformations near morphotropic phase boundary and electrostriction, can be measured. In this chapter, we introduce examples of the application of synchrotron XRD to the time-resolved measurement of the piezoelectric response of ferroelectric thin films.

2 Basic Feature of Piezoelectric Response in Ferroelectric Materials

Ferroelectrics are crystalline materials, which have two or more orientational states with polarization in the absence of an electric field and can be shifted from one to another of these states by an electric field [1]. Since polarization is associated with the relative displacement of cations and anions in the lattice, the lattice will have anisotropic deformation with the elongation along the polarization direction. When an electric field is applied along the spontaneous polarization direction, the lattice is further elongated as shown in Fig. 1a, which is the primary origin of piezoelectric response (more precisely, converse piezoelectric response) in ferroelectric materials. As the electric field induces a change in the d-spacing, this is manifested as a change in peak position in XRD.

Due to the fact that many ferroelectric materials can stabilize the spontaneous polarization along more than one direction (without distinction between plus and minus directions), the ferroelectric materials often have multiple domains where neighbor domains have different polarization directions with an angle other than 180° as shown in Fig. 1b. When an electric field is applied to such materials, the spontaneous polarization, which is not parallel to the electric field, can be directed along with it; this manifested as a change in the 2θ peak intensity. Such change in the polarization direction induces a remarkably large deformation of the material as the difference in the lattice constant along the polarization, and perpendicular to it, is much larger than that of piezoelectric response shown in Fig. 1a. Since this is energetically costly, this typically occurs together with the movement of domain walls which move in order to stabilize/grow domains where the spontaneous polarization

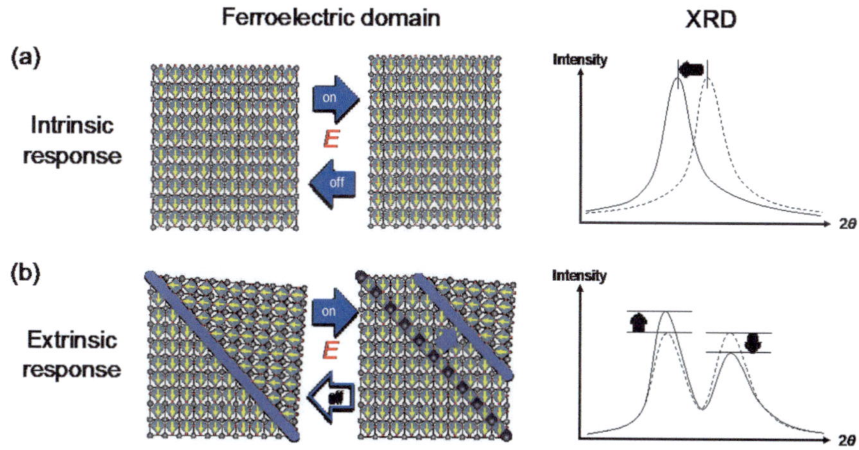

Fig. 1 Two different piezoelectric responses in ferroelectric materials: **a** intrinsic response and **b** extrinsic response

is parallel (or close to parallel) to the electric field. Nevertheless, it is generally believed that the speed of such domain wall motion is much slower than that of normal piezoelectric response without the change in polarization direction.

The former and latter are respectively called intrinsic and extrinsic piezoelectric responses [2]. The intrinsic response provides a fast (below ns) and linear response, and the extrinsic response provides a slower but large displacement. Besides, in the case of thin films, the mechanical substrate clamping poses various impacts on both piezoelectric responses [3–6]. Therefore, their tuning is necessary to meet the requirements for the desired applications. For that, time-resolved measurement of these responses of the materials of interest becomes important.

Time-resolved synchrotron XRD can be a strong tool for such a purpose. When the extrinsic response is not reversible, it can be studied by cumulative ex-situ measurement after applications of short voltage pulses (see Fig. 2) [7]. However, when the extrinsic response is reversible, which is more attractive for most applications, in-situ measurement is required. Hereafter, we focus on in-situ time-resolved synchrotron XRD measurement.

3 Pump and Probe Methods

In order to conduct the in-situ time-resolved synchrotron XRD for the piezoelectric response of ferroelectric thin films, one needs to select an appropriate pump, probe, and synchronization with X-rays, suitable for the purpose of clarification. Their typical combinations are shown in Fig. 3. Here, one should notice that one of those three determines the time resolution for revealing the dynamic behavior of piezoelectric response.

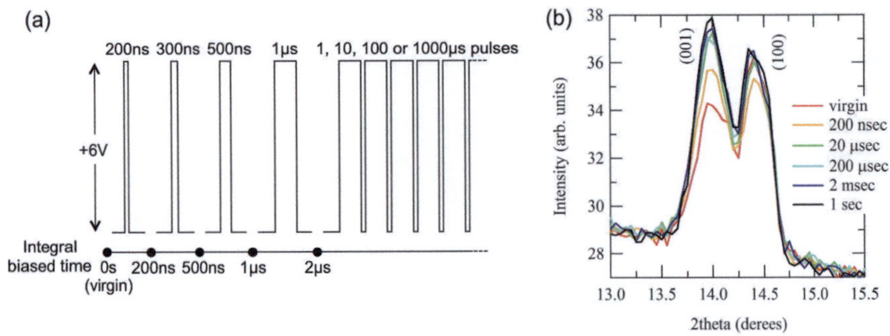

Fig. 2 An example of ex-situ time-resolved XRD study on ferroelectric films. **a** Sequence of pulse voltages applied to ferroelectric $PbZr_{0.3}Ti_{0.7}O_3/PbZr_{0.7}Ti_{0.3}O_3$ bilayer film. **b** $2\theta-\theta$ profiles around the (001)/(100) reflections of the top $PbZr_{0.3}Ti_{0.7}O_3$ layer as a function of the integrated applied pulse width [7]

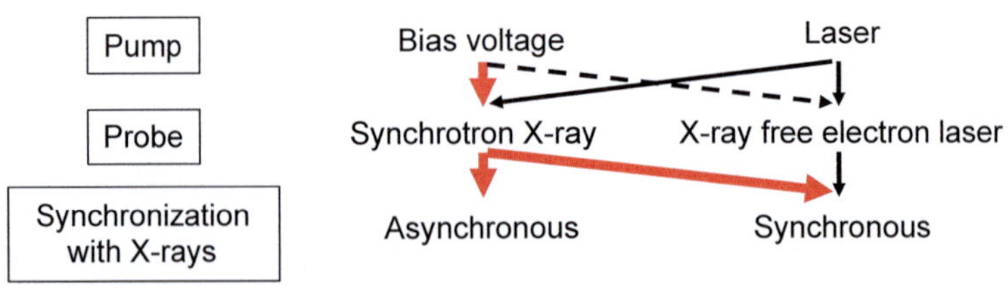

Fig. 3 Combinations of pump, probe and synchronization with X-rays for in-situ time-resolved XRD measurement. The solid arrows show the typical combinations. The bold red arrows are the combinations introduced in this chapter

As for a pump, bias voltage or laser can be used. The bias voltage is used to apply the electric field into the ferroelectric films via the electrodes. This is convenient to integrate with other electrical characterization methods [8]. However, one should be careful about the rise time of the applied electric field due to the RC time constant of the circuit [9]. Since the capacitance of ferroelectric films with electrodes can be a few tens pF or more due to their large dielectric constant (typically, $\varepsilon_r > 100$) and their smaller thickness (typically, a few hundred nm or less), the RC time constant easily becomes 10 ns or more. This means that the dynamics of piezoelectric response can only be discussed for the time scale of the RC time constant or longer. This is not fast enough for the dynamics of the intrinsic response but is enough for the extrinsic response. On the other hand, the electric field can be also applied by using laser. Picosecond, femtosecond or terahertz laser pumping can provide a much shorter time resolution; thus, it is suitable for discussing fast response in the ferroelectric films although the estimation of the actual electric field applied to the film is not straightforward [9–16].

As for a probe, synchrotron X-rays or X-ray free electron laser can be considered. As described in the next section, the synchrotron X-rays are generated from electron bunches passing through the bending magnet, and have an X-ray pulse width of about

40 ps (at SPring-8, Japan in 2002) [17]. Therefore, by synchronizing the detector with the bunches, it is possible to achieve a time resolution as short as the X-ray pulse width. X-ray free electron lasers provide an even shorter pulse width on the order of ~ 10 fs [18, 19]. However, note that these time resolutions are not meaningful when the RC and other time constants and jitter in the measurement system are longer than it. When it is not required to have a time resolution as short as the pulse width of synchrotron X-rays, the synchronization of the detector with X-ray is not necessary. In the following sections, we describe the measurements using the combinations shown by bold red arrows in Fig. 3 as typical examples.

4 Asynchronous and Synchronous Measurements

As explained above, the synchrotron X-rays are generated from electron bunches. There are various bunch modes, and the available modes are different at different storage rings. Nevertheless, they typically have fill patterns with spacings on the order of 10–100 ns. Here, we show two examples at SPring-8 shown in Fig. 4. At SPring-8, the synchrotron storage ring has 2436 buckets for electrons, and the bucket interval is 1.966 ns. The different modes can be designed by filling the electrons, corresponding to the total current of 100 mA, in the desired buckets. Figure 4a shows A-mode with 203 bunches, where the electrons are filled into single buckets with respect to every 12 buckets. Therefore, each single bunch has a current of 0.5 mA and the interval of bunches is 23.6 ns. This mode is useful when the synchronization of the detector with X-ray bunches is not required. Since the X-ray pulses are continuously delivered to the sample at very short intervals, they can be used as "continuous wave (CW)-like" X-rays for most cases. Figure 4b shows H-mode named 11/29-filling + 1 bunch, where 11/29 of full circle, i.e., successive 924 buckets, is filled and a single bunch (5 mA) is set away from the filling. Such a mode, where a single bunch is isolated, is convenient for the synchronous measurements with X-ray bunches.

Now, one should notice that the diffraction intensity is proportional to the acquisition time. The typical photon flux for micro X-ray beam (details described later), at SPring-8 is 10^7–10^{10}/s. If the diffraction intensity per second is 10^3 from the peak of interest and the duration of the single step is 2 ns, the average counts from the single step will be 2×10^{-6}; namely, it is almost zero. Therefore, it is necessary to repeat the voltage pulses until enough intensity is obtained while using the required time resolution.

Let us first introduce how the asynchronous measurement can be designed using the CW-like bunch mode such as 203 bunches shown in Fig. 4a. Figure 5a, b show examples of the time chart for asynchronous measurement using Si avalanche photodiode (APD) detector and single photon counting pixel array detector such as PIxeL ApparaTUs for the SLS (PILATUS) detector, respectively. APD detector has a resolution of a few ns or less, which is shorter than the typical RC time constant of ferroelectric film capacitors. Recall that the observation of dynamic piezoelectric behavior (through changes in 2θ peak position and intensity) is primarily limited by

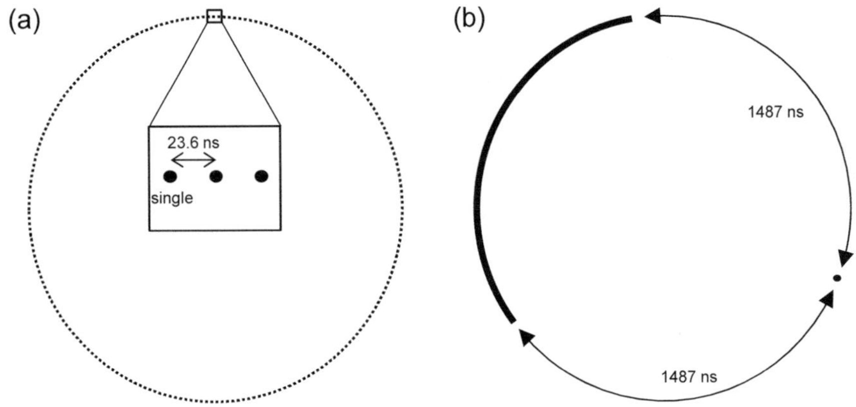

Fig. 4 Examples of bunch mode of synchrotron X-rays at SPring-8. **a** A-mode with 203 bunches, where 203 single bunches (0.5 mA) are set with the same interval. **b** H-mode with 11/29-filling + 1 bunch, where 11/29 of full circle is filled and a single bunch (5 mA) is set away from the filling. Total current is 100 mA regardless of the bunch mode

the RC time constant. A value of a few ns is far slower than the intrinsic piezoelectric response but is faster than the extrinsic piezoelectric response. For this measurement, the periodic signals from the clock built in the delay generator are used as the start of each voltage cycle. Then, transistor-transistor-logic (TTL) signals with a certain delay (ΔT_E) from the clock are used as triggers for the voltage pulses. The start signals and signals from APD are sent to a multi-channel scalar (MCS) to measure the time difference from the nearest start signal. In the case of the PILATUS detector, the minimum exposure time is approximately 300 ns, which is longer than the RC time constant. Thus, the time resolution is limited by the exposure time of the detector. Nevertheless, it is still enough to study the extrinsic response, and there is a remarkable advantage that the PILATUS detector can detect the diffraction for the wide range of reciprocal space at once, which significantly shortens the total measurement time; thus, enabling investigate to the detailed dynamics of extrinsic piezoelectric response. For this measurement, in addition to the TTL trigger signal for the voltage pulses, the TTL signals with a certain delay (ΔT_D) are used as triggers for the PILATUS detector as shown in Fig. 5b. By changing ΔT_D, it is possible to have the set of diffraction data as a function of time.

Next, let us move on to the synchronous measurement using the mode with isolated single bunches such as 11/29-filling + 1 bunch shown in Fig. 4b. Figure 5c shows an example of the time chart for synchronous measurement with the PILATUS detector. For this measurement, specific reference signals from the storage ring are used as the start of each voltage cycle so that the detector can be synchronized with X-ray bunches. In this example, ΔT_D is fixed to be constant during the measurement so that the next single bunch comes during the exposure time of the detector. Since the width of single bunch X-rays on the order of a few tens ps is much shorter than the exposure time of the detector, a time resolution as short as the width of single bunch

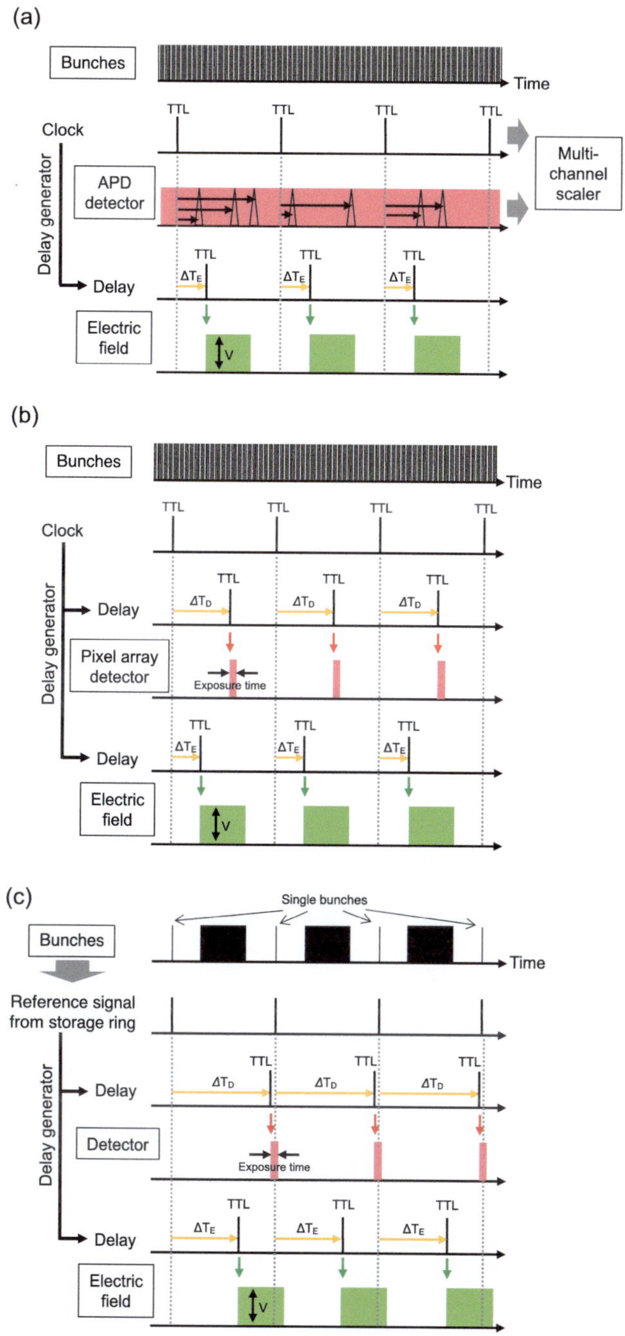

Fig. 5 Schematic time chart for in-situ time-resolved XRD measurement with/without synchronization with X-ray bunches as examples. **a**, **b** Shows the asynchronous measurement using APD detector and PILATUS detector. **c** Shows the synchronous measurement using PILATUS detector

X-rays can be obtained. By changing ΔT_E, it is possible to have the set of diffraction data as a function of time.

5 Experimental Setup, Data Acquisition and Analysis

The experimental setup of time-resolved synchrotron XRD largely depends on the facility and the purpose of the measurement. Figure 6a, b show the photos and the schematic illustration of the setup built at BL13XU and BL15XU beamlines at SPring-8. The energy of the X-rays was selected to be 12.4 or 15 keV by using a double Si(111) monochromator and adjusting the gap of undulator. In order to reduce the RC time constant, an electrode size of few hundred μm or less in diameter is typically used. Therefore, the X-ray beam needs to be focused inside the electrode for the measurement. There are several methods to focus X-rays such as total-reflection mirror, Fresnel zone plate, and refractive lens. All of them can be used with appropriate configurations, but the characteristics of them need to be understood by the users. The total-reflection mirror is achromatic, so the selection of the energy of the X-rays is flexible [20]. On the other hand, unlike Fresnel zone plate and refractive lens, total-reflection mirror changes X-ray beam path; thus, the addition of the total-reflection mirror to the existing diffractometer being used for the other purposes may not be straightforward. Fresnel zone plate offers a minimum focusing size down to sub-micrometer in diameter, but the focusing distance becomes less than that for refractive lens; thus, it needs the consideration of interference with the sample stage and the needle probes for the application of the electric field. The refractive lens offers a focusing size down to a few μm in diameter, which is larger than Fresnel zone plate, but the longer focusing distance (typically 200 mm or more) provides the greater flexibility of design of the sample stage and pump methods. Photon flux through the refractive lens is generally larger than that through Fresnel zone plate. At BL13XU and BL15XU at SPring-8, a compound refractive lens from the Karlsruhe Institute of Technology of Germany [21] was used to focus X-rays inside the electrode in the diameter of 100 or 200 μm. The optimal focusing position was determined by a knife-edge scan before XRD measurement.

The samples are mounted on a translation stage in order to irradiate X-rays inside the electrodes. Before adjusting the X and Y positions, the Z, i.e., the sample height, and the tilt angles along the X and Y are aligned as also done for conventional XRD measurement. For finding the X and Y positions, it is helpful to know the approximate X-ray beam position on the sample on a monitor display. An optical microscope can be used for this purpose. On the basis of the condition that the rotation center of the diffractometer is on the X-ray beam path, it is possible to find the X-ray beam position on the display by rotating the sample stage along φ-axis because the rotation center does not move by the rotation. One can then roughly adjust the X and Y positions on the display so that the electrode can be located at the rotation center. In order to confirm if the X-ray beam is focused inside the electrode as expected, the

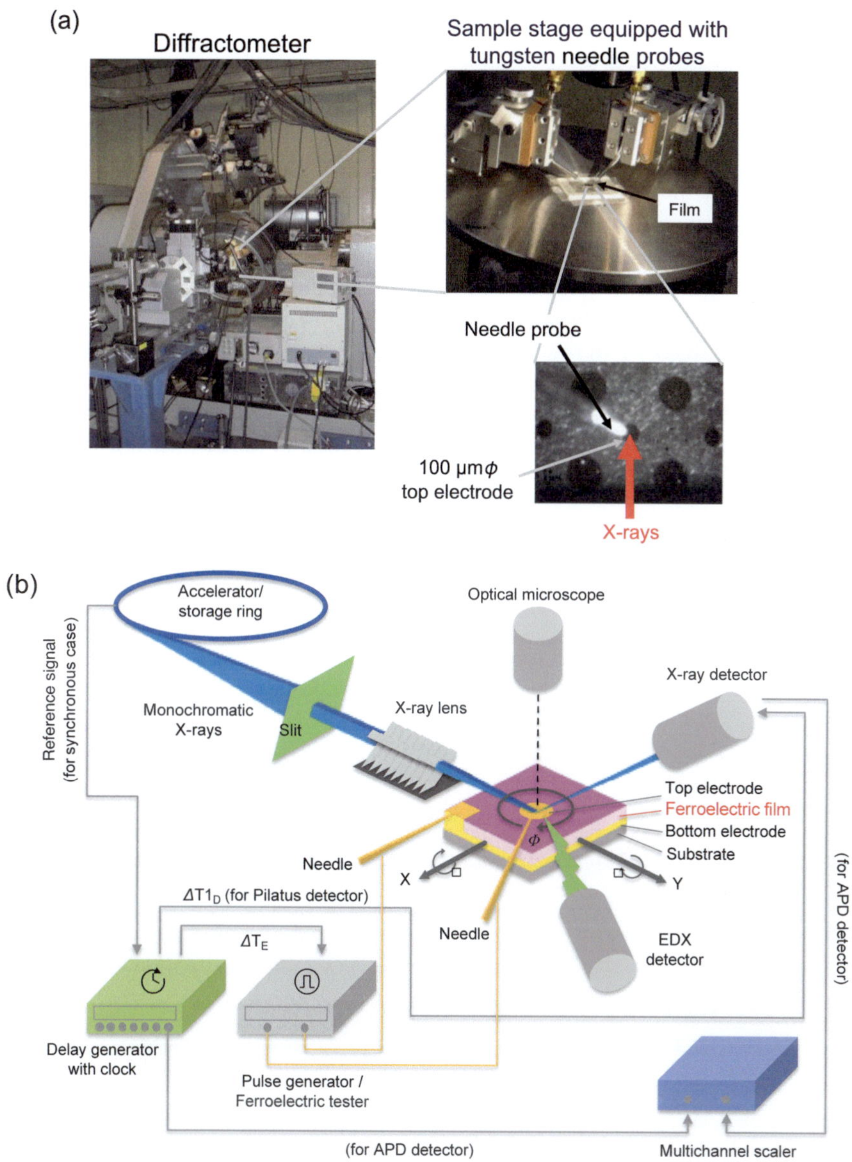

Fig. 6 Photos (**a**) and schematic illustration (**b**) of the setup for in-situ time-resolved XRD measurement built at BL13XU and BL15XU at SPring-8

energy dispersive X-ray spectroscopy (EDX) detector can be used. The exact X and Y positions are determined by X and Y scans of Pt L line across the electrode.

Once the sample position is fixed, the electrical contact with the top and bottom electrodes can be achieved by using the needle probes. One needs to ensure that the electrical property of the ferroelectric film capacitor is as expected, and the films are poled before the measurements. The measurements will be efficiently performed by running the computer control program written for the purpose. For example, one of the widely used software packages for instrument control and data acquisition at synchrotron facilities, SPEC, has a macro function, in which users can easily make the macro loop so that the diffraction angle is changed after the application of several different bias voltages. After the measurement, it is important to check if the electrical property is unchanged as shown in Fig. 7.

The kind of data and its format are different for the used system and setup. Here, we assume that we have the set of data from MCS for different diffraction angles, 2θ as shown in Fig. 7. MCS provides the information of channel and counts. The former can be converted to the time from the start signal, and the latter can be regarded as the diffraction intensity for the total duration of the repeated voltage cycles. Post-treatment using shell scripts and/or in-house programs written in Fortran, C and others can convert and format the data suitable for plotting, and a graphing software having 2D plot function such as gnuplot can visualize the time-resolved XRD under electric field.

6 Case Studies

Hereafter, we introduce three case studies using the above-described time-resolved XRD methods.

6.1 Observation of Intrinsic Piezoelectric Response in Single Domain Pb(Zr,Ti)O$_3$ (PZT) Film [22] and Nanorods [23]

The first case study is the observation of intrinsic piezoelectric response in PZT films. In order to purely reveal the intrinsic piezoelectric response, the polar-axis oriented tetragonal $Pb(Zr_{0.35}Ti_{0.65})O_3$ films with 2100 nm in thickness were fabricated on $CaF_2(100)$ substrate covered with a $SrRuO_3$ bottom electrode layer by using metalorganic chemical vapor deposition. Then, Pt top circular electrodes were fabricated by electron beam evaporation. Since the spontaneous polarization is along the out-of-plane direction, there are no domains having the in-plane polarization component. Figure 8a shows relationship between polarization and electric field of the fabricated $Pt/(001)Pb(Zr_{0.35}Ti_{0.65})O_3/SrRuO_3$ capacitor. Large hysteresis loops originating from ferroelectricity were observed. Once the film is poled with a positive

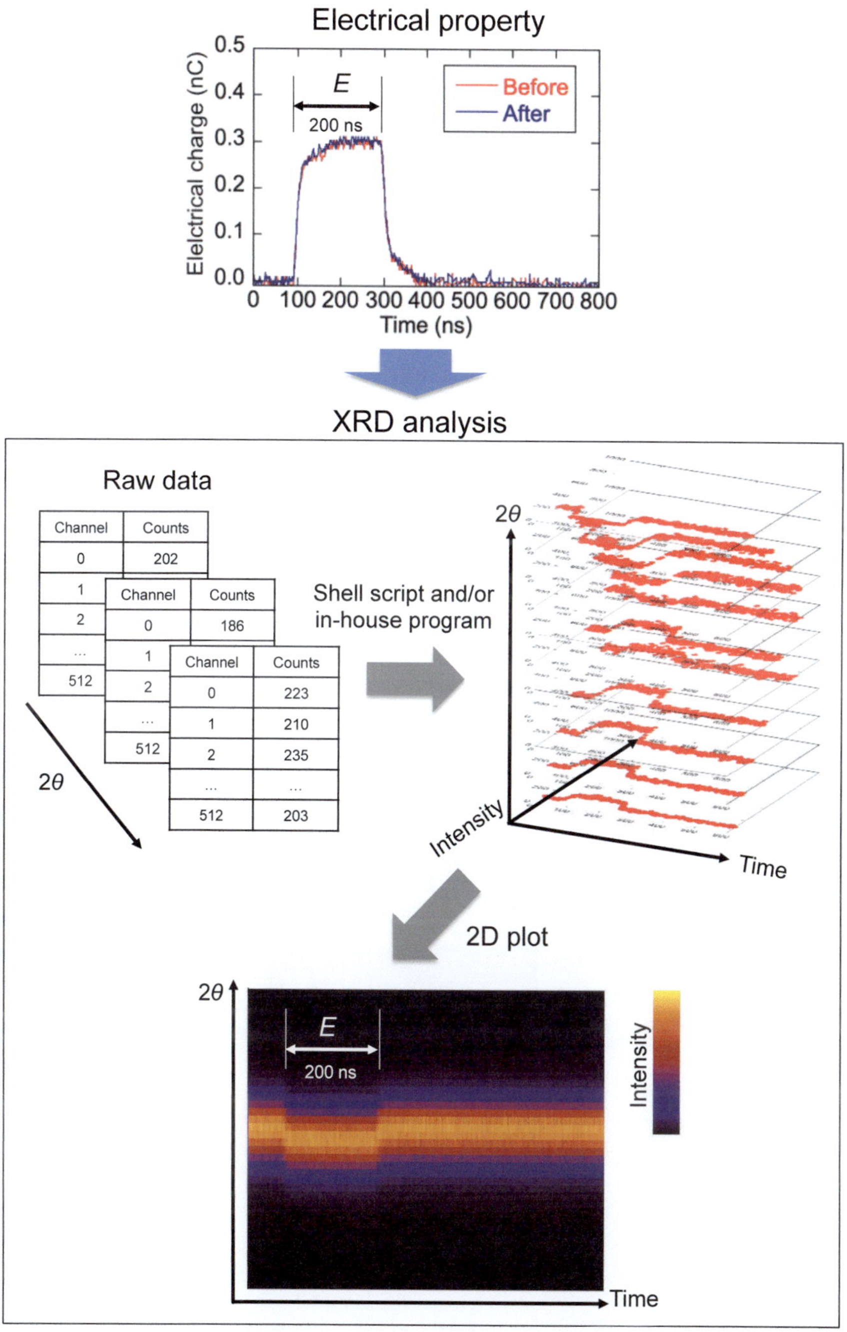

Fig. 7 Schematic diagram for sequence of measurement and analysis. Characterization of electrical property before and after XRD (top) and post-treatment of the data from XRD measurement (bottom)

bias field, application of positive voltage induces the change in polarization strength from P_r to P_{max} (maximum electric polarization at maximum electric field) as seen in Fig. 8a, while the polarization direction is unchanged. In this case, the field-induced lattice strain based on the intrinsic piezoelectric response can only be expected as schematically illustrated in Fig. 8b. Figure 8c shows a change of electric charge of the capacitor with different applied voltage from 10 to 30 V when a rectangular pulse was applied for 100–300 ns in 800 ns total period. Figure 8d shows a time dependency of XRD $2\theta - \theta$ profiles measured under different applied voltages using the asynchronous measurement setup with APD detector. As described above, the dynamic behavior of the intrinsic response is hidden by the RC time constant of the fabricated capacitor structure, but the lattice strain owing to the intrinsic response can be precisely captured. As can be seen in the figure, peak position shifted to a lower 2θ angle under applied voltage and its shift became large with increasing applied voltage. Figure 8e plots the lattice strain estimated from the change in 2θ angle as a function of an applied electric field. A linear relationship was observed between lattice strain and applied electric field, which indicates that the observed lattice strain originates from the piezoelectric response. The observed apparent slope was estimated to be 65 pm/V, which was in the range of the obtained values using piezo force microscopy (PFM) shown in Fig. 8f. As compared with PFM or other conventional methods, the synchrotron time-resolved XRD provides a much more accurate piezoelectric constant due to the high accuracy of the lattice constant measurement.

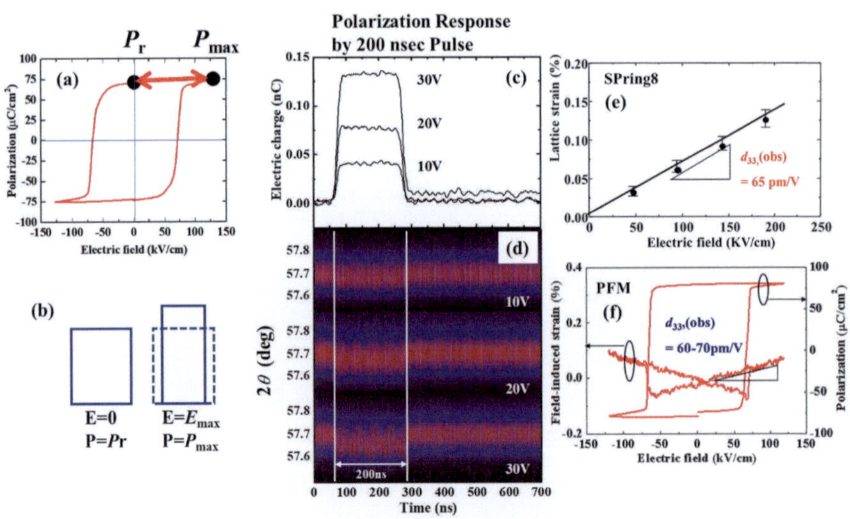

Fig. 8 Ferroelectric and piezoelectric properties of polar-axis oriented $Pb(Zr_{0.35}Ti_{0.65})O_3$ film: **a** relationship between polarization and electric field, **b** expected lattice deformation by electric field, time dependency of **c** electric charge and **d** XRD $2\theta - \theta$ profiles, **e** lattice strain as a function of applied electric field obtained from the data shown in **d**, and **f** electric field dependence of field-induced strain and polarization measured by PFM [22]

The difference in the intrinsic piezoelectric response of the films as compared with bulk single crystals is the influence of substrate clamping [4]. Since the films are mechanically constrained to the substrate, the in-plane lattice deformation by the application of voltage is limited, unlike the illustration in Fig. 8b. Therefore, the observed intrinsic piezoelectric response always is lower than that for unclamped bulk single crystals. This can be solved by fabricating the polar-axis oriented single domain nanorods, where the substrate clamping is negligible. As shown in Fig. 9, the piezoelectric response of nanorods is significantly higher than that of the film and comparable to that for bulk single crystals.

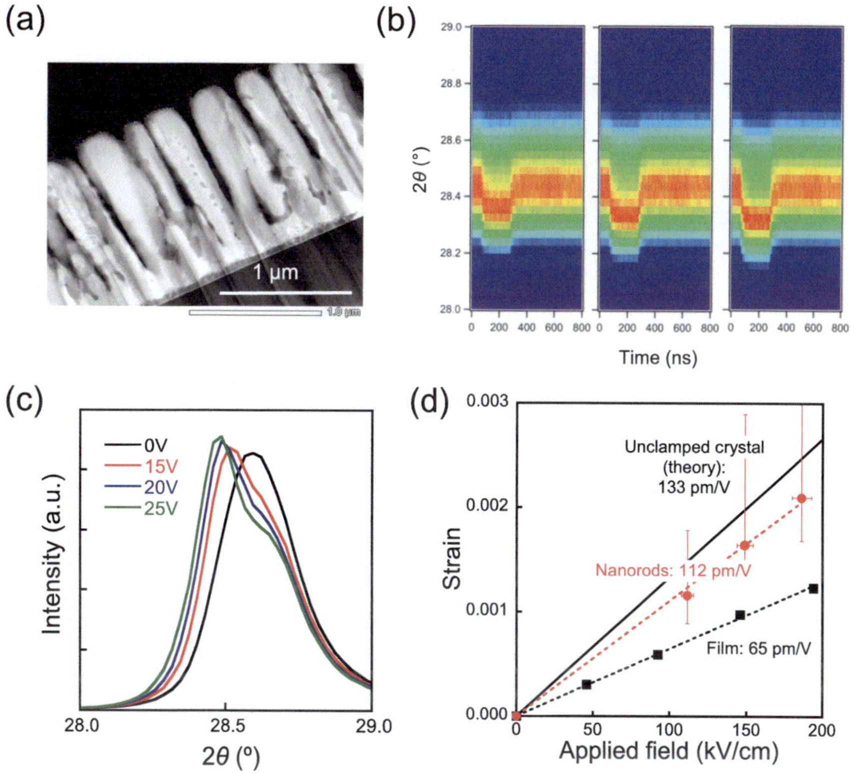

Fig. 9 Structural and piezoelectric properties of polar-axis oriented $Pb(Zr_{0.35}Ti_{0.65})O_3$ nanorods: **a** cross-sectional transmission electron microscopy image, **b** time dependency of XRD 2θ profiles with applications of 200-ns-width voltage pulses, **c** XRD 2θ profiles under different voltages extracted from **b**, and **d** lattice strain as a function of electric field [23]

6.2 Observation of Extrinsic Piezoelectric Response and Its Relaxation Behavior in Polydomain $Pb(Zr_{0.65}Ti_{0.35})O_3$ Film [24]

A more complicated case is the observation of extrinsic piezoelectric response for a (111)/(11$\bar{1}$)-oriented rhombohedral $Pb(Zr_{0.65}Ti_{0.35})O_3$ film. In this film, (111) and (11$\bar{1}$) mixture orientations result in the formation of non-180° domain structure, where the polarization of (111) domains is along the out-of-plane direction and that of the (11$\bar{1}$) domains is tilted as schematically shown in Fig. 10a. Figure 10b shows a time-resolved XRD $2\theta - \theta$ profile for $Pb(Zr_{0.65}Ti_{0.35})O_3$ under the positive electric field by using the same setup used in the first case study. We can find out two things from this XRD $2\theta - \theta$ profile. One is a peak shift, especially for *222* and the other is a relative change of peak intensity between *222* and *22$\bar{2}$*. The time dependencies of lattice strain and volume fraction of (111) orientation defined by $dV_{111}/(V_{111} + V_{11\bar{1}})$ are plotted in Fig. 10c, d, respectively. Figure 10e visualizes the time dependency of *d*(strain)/*dt* obtained from Fig. 10c, d. Interestingly, the response of total elastic strain estimated from both lattice strain and volume fraction change was 80 ns slower than that of the lattice strain estimated from the *222* peak shift, which unveils part of the dynamic behavior of the extrinsic piezoelectric response.

6.3 Synchronous Time-Resolved XRD for PZT Film Using PILATUS Detector [25]

Third case study is the time-resolved XRD system synchronized with X-ray bunches capable of nanosecond characterization of the piezoelectric response of a (111) tetragonal PZT film. An X-ray beam spot with horizontal and vertical dimensions of 1.7 and 0.81 μm, respectively, was achieved with the use of a compound refractive lens. The X-ray pulse was synchronized with an electrical pulse signal and a PILATUS detector (PILATUS 200 K, Dectris, Ltd.) through a delay generator and pulse generator, as explained above (also see Figs. 5c and 6b). The sample was a 750-nm-thick PZT film. The measurement was performed at the BL15XU NIMS beamline, at SPring-8.

We selected the special bunch mode of D with 1/7 filling + 5 bunches in this beamtime. The bunch interval time was 684.3 ns, which was sufficient to obtain an XRD signal from a single bunch. The total current of the electron beam at SPring-8 was 100 mA, which in the D mode was composed of 85 mA for 1/7 filling and 15 mA for 5 bunches. A bunch signal from the D mode was used as the revolution clock signal. Our data in this note were collected from the D mode. Another possible mode was the H mode of the single-bunch flux of 5.0 mA (see Fig. 4b). The X-ray beam was input to a delay generator as a revolution clock signal (Stanford Research Systems, Inc., model DG-645). The signal from the delay generator was input to a pulse generator (Agilent 8114A), and the time of the electric field to the sample was

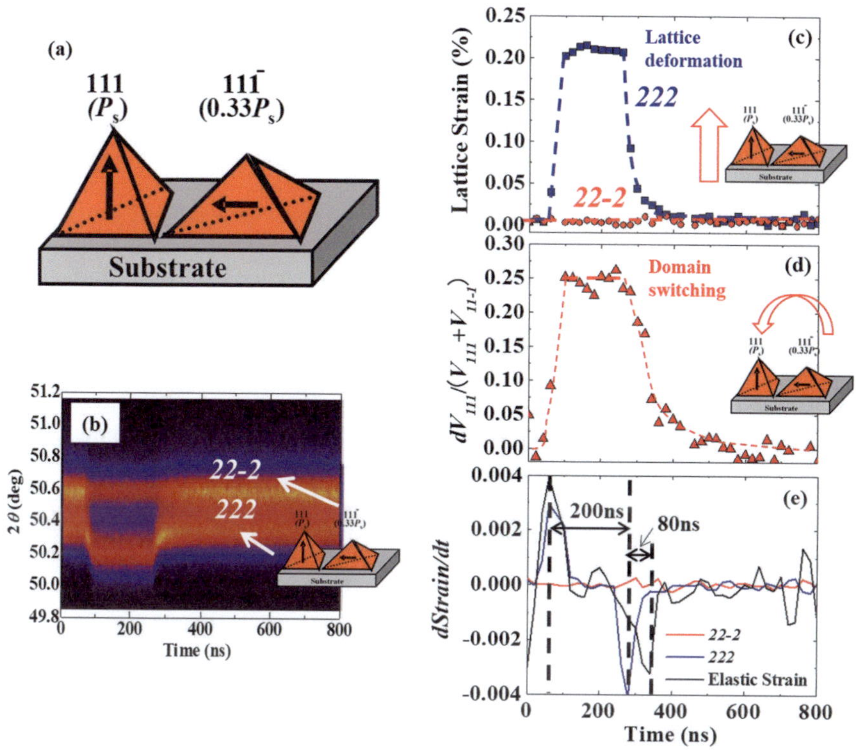

Fig. 10 **a** Schematic model of domain structure of (111)/(11$\bar{1}$)-oriented rhombohedral $Pb(Zr_{0.65}Ti_{0.35})O_3$ film. **b** Time-resolved XRD $2\theta - \theta$ profile under the electric field. Time dependencies of **c** lattice strain and **d** volume fraction of (111) domain defined by $dV_{111}/(V_{111} + V_{11\bar{1}})$, and **e** d(strain)/dt estimated from **c** and **d** [24]

delayed by adjusting the delay generator. The pulse generator can provide a voltage range of 1–50 V and a pulse width of 10 ns–150 ms. The rectangular shape waveform is only available for pulse generator so far. The kinds of waveform can be overcome by introducing other function generator models.

Figure 11a–e show the two-dimensional pixel map of the XRD at PZT 222 Bragg reflection along the out-of-plane direction at different delay times for the application of the electric field, t_0, t_0 + 200, 700, 1100, and 1500 ns (corresponding to ΔT_E in Fig. 5c). We applied 24 V to the electrode with a pulse width of 1000 ns during the unipolar rectangular wave. Since the X-ray pulse width of about 50 ns is shorter than the exposure time of the PILATUS detector, the dwell time of the XRD patterns was the same as the X-ray pulse width. The 2θ peak position of the PZT 222 Bragg reflection was located at 41.32° at V = 0. The Bragg peak position of maximum intensity shifted to a lower angle position of 41.26° and no peak changes were observed along the χ direction at electric field E = 320 kV/cm. Hence, the lattice in the PZT films was only elongated in the out-of-plane [111] direction, which is parallel to the electric field direction. After 1100 ns, the Bragg peak of maximum intensity moved to a higher angle position than the original peak position. This result indicates a change of peak position related to the piezoelectric response. We calculated the piezoelectric constant using the peak position change. Although we did not evaluate the slope of strain value as a function of voltage change, the out-of-plane piezoelectric constant of the thin films was evaluated 51 pm/V using the strain on the electric field of 320 kV/cm. This value is close to but slightly larger than the theoretical value of 49 pm/V expected for the clamped film. Figure 11f shows the 2θ peak position as a function of the delay time. The changes of the peak shape of the PZT films followed the rectangular electric waves with a pulse width of 1000 ns. To find the peak center position of the PILATUS detector, we evaluated the centroid peak position of 2θ of the PZT 222 Bragg reflection, as shown in Fig. 11g. The lattice constant at t_0 was 4.058 Å, as estimated from the peak center position of the PZT 222 Bragg reflection. After $t_0 + 700$ ns, the lattice constant expanded to a maximum of 4.064 Å. The saturation time and recovery time of the peak shift of the PZT films under the electric field were approximately 300 ns. The RC time constant of rise time (10–90%) was described by the RC time constant of $t_{RC} \sim 2.2$ RC, where R is the resistance and C is the capacitance. A resistance of the electrode of 50 Ω was used. The capacitance of PZT films was 1.5×10^{-10} F using the equation of C ~ P_rS/U [26], where P_r is a remanent ferroelectric polarization of the films (46 μC/cm^2) [27], S is the electrode area, U is the voltage pulse height. t_{RC} of PZT films was expected to be 17 ns. The poled (111) tetragonal PZT film has polydomain structure with three equivalent variants of polarization. Here, the vertical polarization components are the same for all the domains, but the direction of in-plane polarization components is different between neighboring domains. The latter might reduce local clamping of PZT lattices by compensating their anisotropic in-plane polar distortion of each other, which is consistent with the observed piezoelectric constants larger than the theoretically expected value for the clamped film [27]. Besides, this might also be related to the longer saturation time and recovery time as compared with the single domain PZT films.

7 Summary

In this chapter, the measurement of the piezoelectric response of ferroelectric thin films was introduced as an example of synchrotron time-resolved XRD. Depending on the time scale of the response of interest, the appropriate selection of pump, probe and synchronization with X-rays is necessary. In addition, the RC and other time constants and jitter must be considered. When the electrode is used to apply an electric field, the X-ray beam needs to be focused inside the electrode. Acquired data can be post-treated using Shell scripts and/or in-house programs for plotting and analysis. We also introduced three case studies using such measurement systems. The basis of the introduced methods can be also applied to reveal the response of other functional materials to the external field.

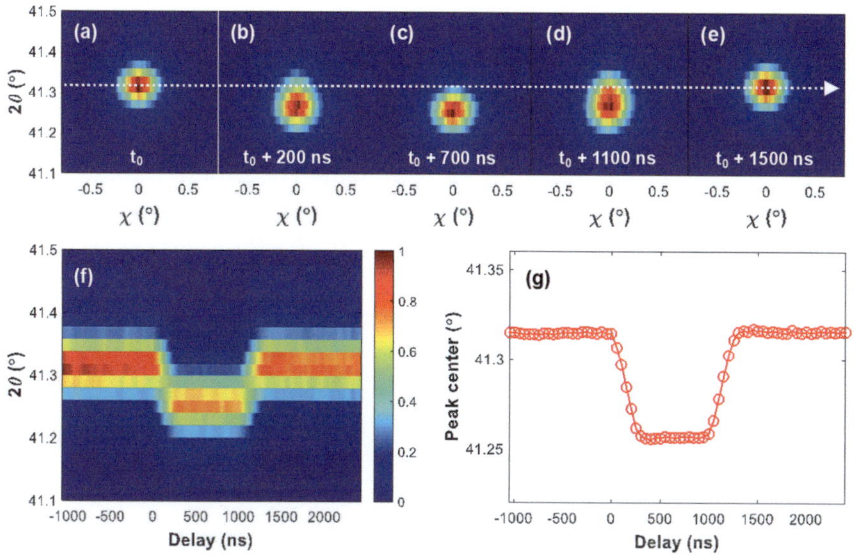

Fig. 11 Two-dimensional pixel information at **a** t_0 and after **b** $t_0 + 200$ ns, **c** $t_0 + 700$ ns, **d** $t_0 + 1100$ ns, and **e** $t_0 + 1500$ ns, respectively. **f** Two-dimensional intensity map and **g** peak center positions of PZT 222 Bragg reflection as a function of the delay time under the unipolar rectangular wave with a pulse width of 1000 ns [25]

Acknowledgements This work was partly supported by the Element Strategy Initiative to Form a Core Research Center, Ministry of Education, Culture, Sports, Science, and Technology of Japan (MEXT) (Grant No. JPMXP0112101001), MEXT Program: Data Creation and Utilization Type Material Research and Development Project (Grant No. JPMXP1122683430), MEXT Initiative to Establish Next-generation Novel Integrated Circuits Centers (X-NICS) (Grant No. JPJ011438), JST-PRESTO (Grant No. JPMJPR16R9), JST-CREST (Grant Nos. JPMJCR20Q2 and JPMJCR24R3), JST Adopting Sustainable Partnerships for Innovative Research Ecosystems (ASPIRE) (Grant No. JPMJAP2312), JST-Adaptable and Seamless Technology transfer Program through Target-driven R&D (A-STEP) (Grant No. JPMJTR24T4), and JSPS KAKENHI (Grant Nos. 18310085, 20510113, 26220907, 26709047, 15H04121, 15K13944, 18H01474, 18K04868, 19H02421, 19H04531, 19J21955, 19K15288, 22H01523, and 22K18295). Time-resolved X-ray diffraction measurements were performed at SPring-8 with the approval of the Japan Synchrotron Radiation Research Institute under Proposal Nos. (Proposal Nos. 2006A1370, 2006B1457, 2005B0189, 2007A1349, 2009A1356, 2011A1550, 2012A1309, 2012A1491, 2012B1034, 2012B1656, 2016B1543, 2017B1574, 2019A1383, 2023A1545, 2024A1778, and 2024B1680) and the National Institute for Materials Science under Proposal Nos. (Proposal Nos. 2014A4908, 2014B4908, 2015A4702, 2017B4903, 2018A4505, 2018A4901, 2018A4905, 2018B4506, 2018B4900, 2018B4907, 2019A4503, 2019B4700, 2019B4905, and 2020A4703). We are also grateful to Dr. K. Yokoyama, Dr. S. Takeda, Professor Y. Kagoshima, and Professor J. Matsui at the University of Hyogo for their technical contributions.

Questions and Problems (Synchrotron Time-Resolved X-Ray Diffraction for Piezoelectric Response of Ferroelectric Thin Films)

Q 7.1. What are the two possible ways in which the piezoelectric response of a ferroelectric under electric field can be manifested in an in-situ XRD experiment? (Hint: see Fig. 1)

Q 7.2.

i. Why is rise time important in in-situ time-resolved XRD characterization of ferroelectric thin films?
ii. Calculate the rise time of a parallel plate ferroelectric capacitor with the following parameters:
    - Area: 1 mm^2
    - Thickness: 10 nm
    - Peak relative dielectric permittivity: 200
    - Resistance: 50 Ω.
iii. Now calculate the same for a capacitor 100 mm^2. What do you learn from this?

Q 7.3. Explain why the authors state "synchrotron time-resolved XRD provides a much more accurate piezoelectric constant due to the high accuracy of the lattice constant measurement."

Q 7.4. Show numerically how the authors arrived at the value of 65 pm/V for the data presented in Fig. 8e.

Q 7.5. Why is clamping by substrate important when considering the piezoelectric response of a thin ferroelectric film?

Q 7.6. Obtain the plot in Fig. 10e using the data given in Fig. 10c, d.

Q 7.7. Numerically obtain the piezoelectric constant using the peak position change for the following dataset: Assume the lattice constant at t_0 was 4.058 Å. After $t_0 + 700$ ns, the lattice constant expanded to a maximum of 4.064 Å. The applied field was 1000 kV/cm.

References

1. M.E. Lines, A.M. Glass, *Principles and Applications of Ferroelectrics and Related Materials* (Oxford University Press, 2001)
2. N. Setter et al., Ferroelectric thin films: review of materials, properties, and applications. J. Appl. Phys. **100**, 051606 (2006)
3. D.-J. Kim, J.-P. Maria, A.I. Kingon, S.K. Streiffer, Evaluation of intrinsic and extrinsic contributions to the piezoelectric properties of $Pb(Zr_{1-x}Ti_x)O_3$ thin films as a function of composition. J. Appl. Phys. **93**, 5568 (2003)

4. K. Lefki, G.J.M. Dormans, Measurement of piezoelectric coefficients of ferroelectric thin films. J. Appl. Phys. **76**, 1764 (1994)
5. P. Muralt, Ferroelectric thin films for micro-sensors and actuators: a review. J. Micromech. Microeng. **10**, 136 (2000)
6. M. Nakajima, H. Nakaki, Y. Ehara, T. Yamada, K. Nishida, T. Yamamoto, M. Osada, H. Funakubo, In situ Raman spectroscopy for characterization of the domain contributions to electrical and piezoelectric responses in $Pb(Zr,Ti)O_3$ films. Appl. Phys. Lett. **97**, 181907 (2010)
7. Y. Ehara, S. Yasui, J. Nagata, D. Kan, V. Anbusathaiah, T. Yamada, O. Sakata, H. Funakubo, V. Nagarajan, Ultrafast switching of ferroelastic nanodomains in bilayered ferroelectric thin films. Appl. Phys. Lett. **99**, 182906 (2011)
8. O. Sakata, S. Yasui, T. Yamada, M. Yabashi, S. Kimura, H. Funakubo, In-situ lattice-strain analysis of a ferroelectric thin film under an applied pulse electric field. AIP Conf. Proc. **1234**, 151 (2010)
9. A. Grigoriev, D.-H. Do, P.G. Evans, B. Adams, E. Landahl, E.M. Dufresne, Synchronizing fast electrically driven phenomena with synchrotron X-ray probes. Rev. Sci. Instrum. **78**, 023105 (2007)
10. T. Ejdrup, H.T. Lemke, K. Haldrup, T.N. Nielsen, D.A. Arms, D.A. Walko, A. Miceli, E.C. Landahl, E.M. Dufresne, M.M. Nielsen, Picosecond time-resolved laser pump/X-ray probe experiments using a gated single-photon-counting area detector. J. Synchrotron Radiat. **16**, 387 (2009)
11. H. Wen et al., Electronic origin of ultrafast photoinduced strain in $BiFeO_3$. Phys. Rev. Lett. **110**, 037601 (2013)
12. Y. Li et al., Giant optical enhancement of strain gradient in ferroelectric $BiFeO_3$ thin films and its physical origin. Sci. Rep. **5**, 16650 (2015)
13. J. Park, Q. Zhang, P. Chen, M.P. Cosgriff, J.A. Tilka, C. Adamo, D.G. Schlom, H. Wen, Y. Zhu, P.G. Evans, Spatially confined low-power optically pumped ultrafast synchrotron X-ray nanodiffraction. Rev. Sci. Instrum. **86**, 083904 (2015)
14. Y. Ahn, J. Park, A. Pateras, M.B. Rich, Q. Zhang, P. Chen, M.H. Yusuf, H. Wen, M. Dawber, P.G. Evans, Photoinduced domain pattern transformation in ferroelectric-dielectric superlattices. Phys. Rev. Lett. **119**, 057601 (2017)
15. Y. Ahn et al., Nanosecond optically induced phase transformation in compressively strained $BiFeO_3$ on $LaAlO_3$. Phys. Rev. Lett. **123**, 045703 (2019)
16. F. Chen et al., Ultrafast terahertz-field-driven ionic response in ferroelectric $BaTiO_3$. Phys. Rev. B **94**, 180104 (2016)
17. M. Yabashi, K. Tamasaku, T. Ishikawa, Measurement of X-ray pulse widths by intensity interferometry. Phys. Rev. Lett. **88**, 244801 (2002)
18. P. Emma et al., First lasing and operation of an Ångstrom-wavelength free-electron laser. Nat. Photon. **4**, 641 (2010)
19. T. Ishikawa et al., A compact X-ray free-electron laser emitting in the sub-Ångström region. Nat. Photon. **6**, 540 (2012)
20. S. Matsuyama, Y. Emi, H. Kino, Y. Kohmura, M. Yabashi, T. Ishikawa, K. Yamauchi, Achromatic and high-resolution full-field X-ray microscopy based on total-reflection mirrors. Opt. Express **23**, 9746 (2015)
21. V. Nazmov, E. Reznikova, A. Last, J. Mohr, V. Saile, R. Simon, M. DiMichiel, X-ray lenses fabricated by LIGA technology. AIP Conf. Proc. **879**, 770 (2007)
22. T. Fujisawa, Y. Ehara, S. Yasui, T. Kamo, T. Yamada, O. Sakata, H. Funakubo, Direct observation of intrinsic piezoelectricity of $Pb(Zr,Ti)O_3$ by time-resolved X-ray diffraction measurement using single-crystalline films. Appl. Phys. Lett. **105**, 012905 (2014)
23. K. Okamoto, T. Yamada, K. Nakamura, H. Takana, O. Sakata, M. Phillips, T. Kiguchi, M. Yoshino, H. Funakubo, T. Nagasaki, Enhanced intrinsic piezoelectric response in (001)-epitaxial single c-domain $Pb(Zr,Ti)O_3$ nanorods. Appl. Phys. Lett. **117**, 042905 (2020)
24. Y. Ehara et al., Ferroelastic domain motion by pulsed electric field in $(111)/(11\bar{1})$ rhombohedral epitaxial $Pb(Zr_{0.65}Ti_{0.35})O_3$ thin films: fast switching and relaxation. Phys. Rev. B **100**, 104116 (2019)

25. O. Seo, J. Kim, C. Song, Y. Katsuya, Y. Shimada, T. Yamada, O. Sakata, Time-resolved X-ray diffraction system for study of $Pb(Zr,Ti)O_3$ films under a temporal electric field at BL15XU, SPring-8. Rev. Sci. Instrum. **90**, 093001 (2019)
26. R.-C. Peng, J.-M. Hu, K. Momeni, J.-J. Wang, L.-Q. Chen, C.-W. Nan, Fast 180° magnetization switching in a strain-mediated multiferroic heterostructure driven by a voltage. Sci. Rep. **6**, 27561 (2016)
27. T. Yamada et al., Negligible substrate clamping effect on piezoelectric response in (111)-epitaxial tetragonal $Pb(Zr,Ti)O_3$ films. J. Appl. Phys. **118**, 072012 (2015)

MIX
Papier aus verantwortungsvollen Quellen
Paper from responsible sources
FSC® C105338

If you have any concerns about our products,
you can contact us on
ProductSafety@springernature.com

In case Publisher is established outside the EU,
the EU authorized representative is:
Springer Nature Customer Service Center GmbH
Europaplatz 3, 69115 Heidelberg, Germany

Printed by Libri Plureos GmbH
in Hamburg, Germany